SPACE
TREK

SPACE TREK

THE ENDLESS MIGRATION

JEROME CLAYTON GLENN • GEORGE S. ROBINSON

Stackpole Books

SPACE TREK: THE ENDLESS MIGRATION
Copyright © 1978 by Jerome Clayton Glenn and George S. Robinson

Published by
STACKPOLE BOOKS
Cameron and Kelker Streets
P.O. Box 1831
Harrisburg, Pa. 17105

Published simultaneously in Don Mills, Ontario, Canada
by Thomas Nelson & Sons, Ltd.

Jacket artwork, © 1977 by The Foundation, Inc. Artist: Rick Sternbach.
The endpiece was adapted from a color painting by Denise Sati © 1977.

Printed in the U.S.A.

Library of Congress Cataloging in Publication Data

Glenn, Jerome Clayton, 1945-
  Space trek.

  Includes index.
  1. Outer space—Exploration. 2. Reusable
space vehicles. 3. Space colonies.
I. Robinson, George S., 1937-   joint author.
II. Title.
TL793.G56 1978 387.8 78-15969
ISBN 0-8117-1581-7

To Thomas Jefferson

Contents

Prologue

The pessimists are wrong.

The planet earth is 5 billion years old, humans 400 million years on the planet, and industrial civilization a latecomer at 200, give or take a few years. Condense that into 24 hours, they say, and modern history begins in the closing minutes just before midnight. It's the end of humanity's day.

Right? Wrong.

It may be true that the earth is 5 billion years old. But add to that the 10 billion years current theories deduce before the sun dies, and the clock changes its reading.

Fifteen billion years reduced proportionately to 24 hours in the life of humanity puts the arrival of modern civilization at 8:00 A.M.—time to wake up and start a new day.

Mankind, for the first time in its history, can now think about surviving for millennia to come. The knowledge and the technology already exist. It has become possible to think in realistic ways about later generations inhabiting the solar system, the stellar system—indeed, the universe.

The space shuttle, the first true space ship, is a *Niña, Pinta,* and *Santa Maria* all in one. It marks the start of an adventurous future more thrilling and evolutionarily more important than anyone can imagine. The present decade brings the dawning not only of the technology that can penetrate outer space, but also of the new intelligence that is making the first imaginative steps into the universe.

Earthkind's expansion into the reaches beyond the planet is not just the product of science fiction's fertile imagination. Nor is it the expensive divertissement of the few. As the embodiment of "whither mankind," it is everybody's business in this, the sole stopping place before the stars.

Space Trek tries to demonstrate clearly just how much space travel involves everyone on earth—now and into the far future. It is the authors' attempt to sound the 8:00 A.M. wake-up bell—time to get moving in the start of humanity's new day.

GEORGE S. ROBINSON
JEROME CLAYTON GLENN

Acknowledgments

Dorothy C. Glenn, Ann K. Robinson, Pat Pastor, Dick Hoagland, Barbara Hubbard, John Whiteside, William Brown, Herman Kahn, Thomas Turner, Carolyn and Keith Henson, Wilton Dillon, Check Hewitt, Allan Ladwig, Shaun Murphy, John Yardley, Henry Kolm, Stan Rosen, Pat Daley, Charles Sheldon, Ursula Kruz-Vacienne, Kai Lee, John Thomas, and Joyce Russell for their help in the preparation of this book. Special acknowledgment to Carolyn Amundson for very practical assistance, constructive additions, and abiding faith in the manuscript and its purpose.

CHAPTER 1

Shuttle to Space:
The Beginning of Migration

*All this world is heavy with the promise of greater
things, and a day will come, one day in the unending
succession of days, when beings, beings who are now
latent in our thoughts and hidden in our loins, shall
stand upon this earth as one stands upon a footstool,
and laugh and reach out their hands amidst the stars.*
—H.G. Wells, 1903

Earth's first space ship has arrived.

Christened the *Enterprise*, it makes the vision of H.G. Wells
more than fantastic imagination. The space shuttle begins to
make it achievable fact.

The habitation of the universe by the residents of Earth is
possible within as little as twenty years. Humanity has begun
making plans to "laugh and reach out their hands amidst the
stars."

NASA's shuttle marks the start of the second age of space
flight. As Arthur C. Clarke put it, "The shuttle is to space flight
what Lindbergh was to commercial aviation."

Administratively, the shuttle is a godsend. It ties together
nicely the many interests in space. Both people and satellites
can be carried to space in the shuttle.

NASA has identified a wide range and large number of

Close-up of space shuttle *Enterprise* and her first test pilots, Fred Haise and Gordon Fullerton (dark suits). *Courtesy of NASA*

feasible space activities that help people right here on earth. These are satellites that work in the areas of communications, navigation, resource and pollution location, weather warning, and crop prediction. But in addition to the already present and practical satellites, NASA studies have emphasized the need for humans in space; it has even considered space tourism as a not-too-distant reality. There is an essential integration and interdependence of the unmanned and manned programs in terms of economic, engineering, and scientific benefits.

NASA management also has noted rather emphatically the driving curiosity and "need to know" of humankind which is sufficient reason in itself for the manned portion. For a Federal agency whose programs were, and continue to be, under attack as nonessential in the face of compelling social needs and demands for public funds, this is a heady and daring assertion of the importance of our individual and collective spiritual needs.

The Skylab missions reflected successful integration of manned and unmanned space objectives. Even after the 84-day

Skylab mission, there were no insurmountable psychological and physiological problems, a fact which had an influential bearing on NASA's immediate space exploration planning. NASA put the major portion of its efforts and funding into a human space transportation system, designed around the development of a relatively inexpensive, reusable space shuttle.

The Space Shuttle Solution

Until now, the guiding principle in NASA operations has been to build spacecraft as simple, uncomplicated, and reliable as possible. This has required a design approach that keeps all operational complexities on the ground wherever possible for easy personnel access. But, as observed by Captain Robert F. Frietag, Deputy Director of Advanced Programs in NASA's Office of Space Flight:

> With the achievement of easy access to space and ultimately permanent occupancy, the opportunity presents itself to reverse this process and to develop much larger, more complicated satellites and greatly simplified ground stations.

Space shuttle can be used for such special missions as placing a booster on the Skylab to lift it to a higher orbit. By preventing Skylab IV from falling back to Earth, we may have preserved the first space museum in orbit. *Courtesy of NASA*

"Mars Across the Sand River" looks much like the "iron horse" or chariot of the future. *Painting by Andrey Sokolov, Courtesy of the Smithsonian Institution*

With this kind of practical reasoning in mind, NASA management has committed itself in large part to the development of manned space transportation and the construction of outer space facilities necessary for long-duration, and ultimately permanent, human occupation of near and outer space.

John F. Yardley, NASA's Associate Administrator in the Office of Space Flight, compares the benefits of developing an economical space transportation system to those which flowed from the improvement of transportation systems throughout history. It is the "iron horse," the covered wagon, the sailboat, and the chariot of the past. The principal benefit historically has been a sharp rise in the standard of living for people living in areas affected by the newly successful means of transportation. The current space transportation system is being developed basically to lower the cost of space activities and to "provide the flexibility required for present and future payloads."

In 1973 a NASA-sponsored study showed that the cost of developing the space shuttle system would be returned (the so-called "payback period") in twelve years. From then on it is a profit to the government and the national economy. This figure assumes at least twenty-five flights a year and a shuttle cost of $14 billion. But it does *not* include such an important factor as rampant inflation.

In the present atmosphere of Federal expenditure justifications, in terms of cost-effect analysis and quantifiable public benefit, one might suspect this type of reasoning is a necessary component of the Federal appropriations game. One might also wonder, though, what would have happened if Thomas Jefferson had been convinced or motivated by this type of justification for the Lewis and Clark expedition. Pretty dry stuff and not much personal or public inspiration! But Yardley firmly believes the shuttle will give the United States a ten- to twenty-

The full space shuttle configuration, complete with large external liquid oxygen tank and two extending solid fuel tanks. The liquid oxygen tank was not originally designed to be recovered, but current plans are to use it in space as living quarters or for other purposes. The solid fuel tanks are recoverable and reusable.

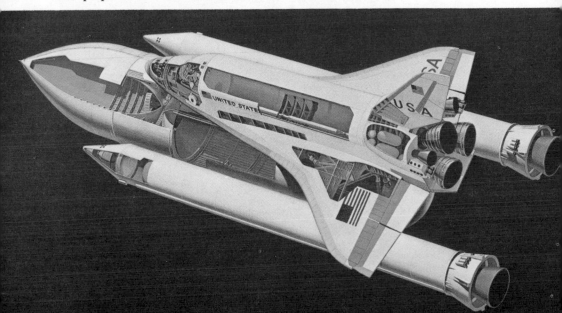

year advantage in space exploration and development. He is convinced NASA's future hinges on its success.

Realistic Flightpath to the Exotic

Regardless of political or budgetary considerations, NASA has undertaken a practical approach to space exploration and development, with a view to more exotic, long-range objectives as well as those which are immediate and practical. The space transportation system will include a space shuttle orbiter, an expendable external tank containing fuel for the orbiter's main engines during ascent to orbit, and two booster rockets which are recoverable and reusable. Currently the main external fuel tank is expendable; however, several studies are underway to find ways to reuse it. One possibility being considered even now is to use the external tank for temporary housing in space.

At present, NASA's goal is a two-week turnaround time on the ground for reuse of the shuttle orbiter, or 160 hours of work actually required after it returns from a mission. Not only will new payloads have to be installed, but also inspection and any necessary repair must be made of the thermal protection system, main and auxiliary propulsion systems, power units, flight instrumentation, and communications system.

The basic characteristics of the shuttle system are:

Length
> System—184 ft., or 56m.
> Orbiter—122 ft. or 37m.

Height
> System—76 ft., or 23m.
> Orbiter—57 ft., or 17m.

Wingspan
> Orbiter—78 ft., or 24m.

Weight
> Gross lift-off—4,500,000 lbs., or 2,000,000 kg.
> Orbiter landing—187,000 lbs., or 85,000 kg.

Cargo Bay
    Orbiter—60 ft. long, or 18m.
           —15 ft. wide, or 5m.

At the forward end of the orbiter is a two-level cabin for the crew and other passengers. From the upper-level flight deck, the crew is responsible for the launching, maneuvering in orbit, re-entry, and landing. Seating and living areas for the passengers are in the lower deck.

Passengers and members of the crew will be subjected to a gravity load of only 3g during launching, and only 1.5g during re-entry. These minimal physical effects, as well as an ambience of one sea-level atmospheric pressure, will permit passengers to travel with much less than the physical perfection required of the current corps of highly trained astronauts.

The primary mission of the shuttle as presently conceived is basically simple. It is to place a variety of satellites in Earth orbit, maintain them, and return them to Earth for refurbishment and reuse. But if the manufacturing of solar power satellites becomes national policy, a greatly expanded program is inevitable. In either case, each shuttle mission will remain in orbit anywhere from fewer than seven days to a month.

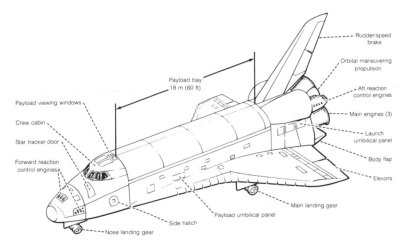

Space shuttle orbiter. *Courtesy of NASA*

Artist's view of flight crew working controls while satellite is placed in orbit. "Vertical" beds for weightless sleeping are shown below. *Courtesy of NASA*

The projection for the 1980s is approximately sixty shuttle flights a year. By that time, it is assumed that the shuttle will be involved in performing tasks for both public and private interests—the government, international organizations, industry, the academic and scientific communities. Shuttle service will be supplied to the Department of Defense on a priority basis; that arrangement is built into the present plans. But if Soviet military space activity continues at its present rate, it is probable that the U.S. military will seek to establish its own operational orbiter capability independent of NASA.

The so-called upper stages of the space transportation system will permit the deployment of satellites to higher orbits than the shuttle orbiter can attain alone. According to John Yardley, "tentative agreement with COMSAT and Ford-Aeroneutronics for INTELSAT-5 launches has been made" that includes operating procedures for the upper stage system. A flight to demonstrate certain uses of this system is planned for as early as 1979.

The "Getaway Special"—From AMTRAK to SPACETRAK

The first reusable shuttle has already been "rolled out" and flight-tested in Earth's atmosphere. Technically referred to as Orbiter 101, former President Gerald Ford christened it the *Enterprise*. More about the naming later.

Routine human space flight has become a reality. These leaps from Earth—"Getaway Specials" if there ever were any—have created enough interest even before they start so that seventy-five space shuttle flights have already been reserved. The British Aircraft Corporation wants to act as international broker for the sale of shuttle space; they have reported Red Chi-

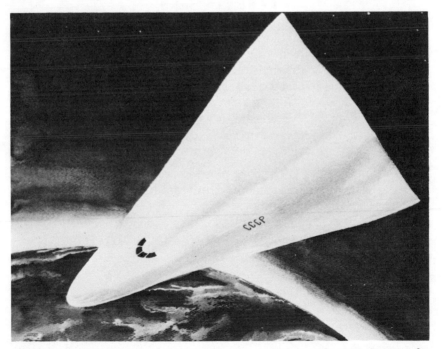

Artist's conception of the Soviet "shuttle" drop-tested in early 1978. *Painting by George Robinson; Photographer, Carolyn Amundson*

na's interest in the $10,000 special. (China has been told to contact NASA and the State Department directly.)

Along with the European Space Agency's use of the shuttle, West Germany has requested two additional shuttle flights of its own. The Communications Satellite Corporation has paid $100 thousand down for a complete shuttle flight to launch INTELSTAT-5. The shuttle prices are fixed up to 1983, but they are related to the Department of Labor's projected price index and may change.

The following NASA chart lists customer, recipient, and number of payloads reserved so far on the space shuttle, in order of payment. First to pay have first choice of flight. Individuals have not been included on the list of industrial users.

SMALL SELF-CONTAINED PAYLOADS

CURRENT USERS

		PAYLOADS
Earnest Money Received For:		**211**
Industrial —	Pharmaceutical Research,	
	Materials Testing and Processing	
	Foreign 35	
	Domestic 91	
	Total **126**	
Education —	Pure and Applied Research by	
	Graduate and Undergraduate	
	Students	
	Foreign 0	
	Domestic 43½	
	Total **43½**	
Individual —	Private Citizens Testing	
	Innovations and Research	
	Ideas in Space	
	Foreign 10	
	Domestic 31½	
	Total **41½**	

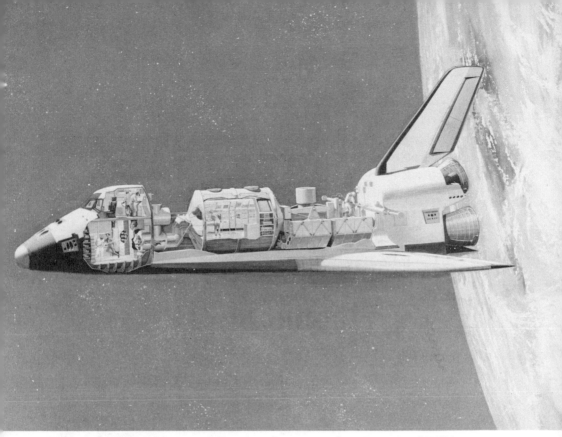

Cutaway showing multipurpose lab in the orbiter's cargo bay. *Courtesy of NASA*

SMALL SELF-CONTAINED PAYLOADS

INDUSTRIAL USERS

1. Battelle
2. Johnson and Johnson
3. Christian Rovsing — Denmark
4. British Aircraft Corporation
5. Erno
6. Versa-Steel, Inc.
7. The Downey Group
8. Marshall-McShane
9. JMSA Flight Safety Products and Services
10. Automation Industries, Inc.
11. Columbia Pictures
12. Columbia Pictures (Steven Spielberg of *Close Encounters of the Third Kind*)
13. Dow Chemical U.S.A.
14. Katy Industries, Inc.
15. Canadian Engineering Publications Limited

16. General Electric Space Division
17. G. S. Roberts Foundation
18. Ford Motor Company
19. Quest 77 Magazine
20. Metrocom, Inc.
21. World Energy Systems
22. Houlihan, Lokey, Howard & Zukin
23. Magnetic Controls Company
24. McDonnell Douglas Astronautics Company
25. The Yomiuri Shimbun — Japan
26. Intermetrics, Inc.
27. Edsyn, Inc.
28. International Technical Associates
29. The Hazard Company
30. Northrop Services, Inc.
31. Second Source Industries — Division of I Corp.
32. Odetics, Inc.
33. Surge, Inc.
34. Quality Chekd Dairy Products Association
35. Austron, Inc.
36. DFVLR-BPT —Federal Republic of Germany
37. Rockwell International Space Division
38. International Communications Management, Inc.
39. Janos Beny Innovations
40. Eastman Whipstock
41. Norment & Castleberry Insurance Company
42. Pullman Standard
43. Jenkins Publishing Company
44. Liller Neal Battle & Lindsey, Inc.
45. R. Johns, Ltd.
46. Secon Metals Corporation
47. Mission Research Corporation
48. General Dynamics Convair Division
49. International Harvester
50. Jay Robinson Aviation & Aerospace
51. Cypress-Caddo Corporation
52. Trinco, Inc.
53. Tramway Museum Society — England
54. Starlog Publications
55. Electrochemical Supplies, Inc.
56. Life Systems, Inc.
57. Astro—Arc Company
58. Soil Systems, Inc.

SMALL SELF-CONTAINED PAYLOADS
EDUCATION USERS

1. Gilbert Moore
   Utah Citizen for:

   Utah State University (½)
   Weber State College and Weber
     County School District, Utah
   Box Elder County School District
     and Ogden City School District,
     Utah
   Educational Institute - TBD (2)

2. Rex Megill
   Professor, Utah State Univ.
     for:

   Utah State University

3. Edward Buckbee
   Alabama Space & Rocket
     Center for:

   University of Alabama (¼)
   Auburn University (¼)
   Alabama A&M University (¼)
   University of Alabama
     in Huntsville (¼)
   Educational Institute - TBD

4. James Kordig
   Utah Section of AIAA for:

   University of Utah

5. Harold Ritchey
   Utah Citizen for:

   Purdue University

6. Marvin Mathews
   Space Center Rotary Club
     for:

   Clear Creek Independent
     School District,
     Houston, Texas

7. Donald Myronuk
   Professor for:

   San Jose State University

8. C. A. Jacobson
   McDonnell Douglas
     Technical Services Co.
   Houston Astronautics
     Division for:

   Prairie View A&M University

9. John Wittry
   Professor for:

   USAF Academy

10. J.F. Niebla
    Orange County Chapter   University of California
       AIAA for: at Irvine

11. Brand Griffin   Rice University School of
    Professor for: Architecture

12. Garland Peed III   San Diego Community
    Chancellor for: College District

13. Richard Azar II   University of Texas at El Paso
    Coors Dickshire, Inc., for: El Paso Independent School
                                     District & Ysleta Independent
                                     School District
                                  Joint El Paso/Juarez Research
                                     Project

14. Robert M. Howe   University of Michigan — Dept.
    Professor for: of Aerospace Engineering

15. Fred A. Wulff for: Goddard Explorer Post 1275

16. Kinnaird R. McKee
    Superintendent for: U. S. Naval Academy

17. Thomas B. Murtagh
    AIAA—Houston Section
       for: Rice University
18. John G. Adams
    Adams Extract Company,
       Inc., for   Southwest Research Institute

19. Walter L. Fix for: Shepherd College and
                                  Roanoke College

20. Theodore J. Rosenberg
    Professor for: University of Maryland

21. M. C. Rasmussen
    University of Texas Alumni
    at McDonnell Douglas
    Technical Services Com-
       pany, Inc., for: University of Texas at Austin

22. Jay Burns
    Professor for:

        Florida Institute of
        Technology

23. John P. Castleman, Jr., for:

        Williams College,
          Williamstown, Ma.
        Woodberry Forest School,
          Woodberry Forest, Va.
        Highland Park High School,
          Dallas, Tx.

24. Thomas Teichner
    General Steel Fabricators,
      Inc., for:

        Niskayuna, N.Y., High School
          Science Dept.

25. Larry L. Gasner
    Professor for:

        University of Maryland

26. David Shore
    RCA for:

        Camden High School and
          Woodrow Wilson High School
          Camden, N.J.

27. Dr. R. F. Brodsky
    Professor for:

        Iowa State University of
          Science and Technology

28. Dr. Madsen Pirie
    Professor for:

        Adam Smith University

29. Dr. P. R. Smith
    Professor for:

        New Mexico State University

30. Terry Borton

        Xerox Education Publications

31. F. B. Gillespie

        Alumni Association for
          Lafayette College

32. George W. Scott

        Los Angeles Section AIAA for
          Education Institute - TBD

33. Mark K. Craig

        Houston Section AIAA for
          Texas A&M University

SMALL SELF-CONTAINED PAYLOADS
INDIVIDUAL USERS

1. Reiner Klett — German Consultant

2. Frank Lenahan — California

3. Gene McCoy — Florida

4. Gilbert Moore — ½ for Personal Use (½ to Utah State Univ.)

5. Thomas Hanes — California

6. Richard Mastronardi — Massachusetts

7. Shirley Arnowitz
   Ann Pincus — Maryland

8. James O. Matzenauer
   R. G. Canetti — California

9. Mike Watson — Arizona

10. Stanton L. Eilenberg — California

11. Jack J. Gottlieb — Maryland

12. Mel Chaskin — Virginia

13. Daniel B. Anderson — California

14. David Brinegar — Nebraska

15. Anton K. Simson — California

16. Anand Prakash — Washington, D.C.

17. Robert Macauley — Connecticut

18. E. G. Wolf   Jordan/Tamraz/Caruso
                                    Advertising for Client
                                    Rich Port, Illinois

19. Robert L. Staehle   California

20. Earl R. Nadeau   New Jersey
    Nicole Nadeau
    Christopher Nadeau

Spacelab—Europe's Contribution

The star payload of the shuttle is the Spacelab, which is being developed by the European Space Agency. This is the largest single project involving international cooperation in space. It is designed to provide an orbiting laboratory for scientists and technicians to perform experiments and carry out other tasks, much in the same manner as the highly successful American Skylab and Soviet Cosmodome missions. But the differences are important.

The Spacelab, to be carried in the cargo bay of the shuttle, will be reusable. It can be flown in many different configura-

A small experimental space station can be placed in orbit and brought back to Earth by the shuttle. The Canadian manipulator arm picks up or lifts out the space luggage. *Courtesy of NASA*

tions, therefore permitting a wide variety of tasks and experiments to be undertaken from one mission to the next. The flexibility built into the Spacelab makes all this relatively inexpensive.

Margaret Mead and many others believe space may well be the best cultural medium to bring a constantly warring world peacefully together. Spacelab presents an early example of such possible future international cooperation. Ten nations from the European space community have agreed to commit approximately $500 million to design and deliver one Spacelab unit to the United States. West Germany, Italy, France, the United Kingdom, Belgium, Spain, the Netherlands, Denmark, Switzerland, and Austria make up that international group. The United States understands that this commitment marks an unprecedented cooperative undertaking.

Rescue, Crew Training, Tourists, and Women

The orbiter can carry up to seven crew members into orbit, as well as the payload. The shuttle system can be readied for

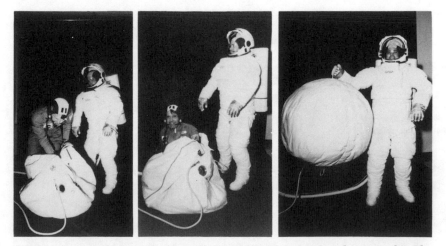

Just in case someone has a problem up there, NASA has developed a rescue kit. The flight crew gets space suits, but the shirt-sleeve environment allows passengers to skip them. But should a problem occur, each is supplied with this "Rescue Embryo."
Courtesy of NASA

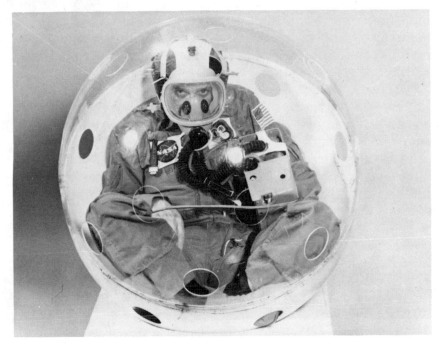

Courtesy of NASA

rescue launch and operations with twenty-four hours of
standby status and can accommodate as many as ten people,
including the three-man rescue crew. Fifty people could be
seated airplane-style in a somewhat modified cargo bay when
the era of space tourism arrives.

Crew training for the space transportation system is
underway. However, the present astronaut group of twenty
pilots and nine scientists is insufficient to handle the planned
schedule of sixty flights per year. Yardley estimates that by the
year 1999, two thousand people will have gained invaluable
experience as participants on shuttle flights, along with
perhaps one hundred "tourists" who have the necessary
$100,000 for a ticket.

NASA has announced employment opportunities for
additional pilots, as well as mission and payload specialists.
Nichelle Nicols, who played the communications officer in
Star Trek, was hired by NASA to recruit women and minorities

Four-person crew seated in launch configuration on the upper flight deck. *Courtesy of NASA*

Women Mission Specialists/Astronaut candidates from left to right: Margaret R. (Rhea) Sheddon, Anna L. Fisher, Judith A. Resnik, Shannon W. Lucid, Sally K. Ride, and Kathryn D. Sullivan. *Courtesy of NASA*

Astronaut Candidates Guion S. (Guy) Bluford (seated), Ronald E. McNair, left, and Frederick D. Gregory. *Courtesy of NASA*

and change the male-dominated character of the current United States space program. The selection and appointment of candidates was made in early 1978.

Permanent Occupancy—Limited Self-Sufficiency

John Yardley has stated that NASA's principal long-term manned space objective is the "achievement of permanent occupancy and limited self-sufficiency in space." The shuttle can be thought of as an industrial sperm cell toward that end. The next step, Yardley believes, will be to construct a permanently occupied near-space station which "would provide the opportunity to establish bases where fabrication, erection, and deployment of large structures for a great number of new missions could take place."

This space construction base concept is now receiving

Artist's conception of the launch of a large payload by a heavy-lift launch vehicle. *Courtesy of NASA*

OTRAG of Germany has a launch vehicle designed to compete for shuttle business.
Courtesy of Frank K. Wukasch, OTRAG, Stuttgart, Germany

primary emphasis in NASA's long-range planning. At this point, there is no technological limit to the size of structures that can be erected in space, and NASA is presently studying the design requirements of a heavy lift launch vehicle which would permit the placement of very large payloads into near-earth orbit.

A businesslike approach to the development of industry in space is integral to NASA's space construction base planning. The objective is to "produce a saleable/profitable product or service which companies [or individuals] are willing to pay for." Unfortunately, much of this concept of space industrialization envisages NASA as the on-site supplier of primary services, if not products. Yet it seems only reasonable to assume that at some point NASA's primary role will end, and development by private industry will begin. That has been the route traditionally followed by almost all activity originally government-sponsored in the United States. Taking it even further, perhaps industry should be encouraged to develop its own space transportation capabilities—that's in addition to the building and habitation of space construction bases and permanent manufacturing communities (a new variation of the "company town").

Toward this end, various early efforts are being made by aerospace organizations to involve themselves in commercial launchings. Certain American companies have approached French authorities with plans for commercial launch sites at the equator—so far without success. A West German firm (OTRAG) has entered into an agreement with the government of Zaire to undertake suborbital tests of that company's commercial launch vehicle. Once private industry gets involved, it seems likely that the shuttle will also use non-American ports.

Even in the face of competition from the space shuttle, plans are underway for alternative commercial launching options and capabilities.

The European Space Agency's three-stage rocket, *Ariane*, has ten European backers, with France paying the most. With its first test scheduled for mid-1979, it is expected to be competitive with the space shuttle by 1981. It will launch geostationary satellites twenty-two thousand miles up from Kourou, French Guiana. The American space shuttle goes to a much lower orbit, but with the interim upper stage (IUS) booster, it also can put satellites up into higher orbit. *Ariane's* launch cost will be about $20 to $25 million, making it quite competitive with the space shuttle.

Japan, in full cooperation with the United States, is testing what it calls the "N" vehicle. Like the *Ariane*, the "N" vehicle will go into operation in 1981, launching geostationary satellites. The Space Cooperation Agreement between the United States and Japan allows American firms to help Japan develop this rocket, but limits the use of the rocket only to Japanese payloads and maintains that any satellites from other countries

Looking back to 1938, when the British flew piggyback, too. *Courtesy of NASA*

The test space ship *Enterprise* loaded on a 747 jet with the greatest airplane of its day—the DC-3—looking on from the foreground. *Courtesy of NASA*

The *Enterprise* prior to takeoff on its historic test flight, October 12, 1977. *Courtesy of NASA*

The *Enterprise* breaks free, signaling the beginning of the Second Age of Space. *Courtesy of NASA*

can be launched only with prior American approval. In this way, the United States can control Japanese competition with the space shuttle.

The Sabre Foundation, a private operating foundation which provides financial assistance for charitable, scientific, and educational purposes, is sponsoring a study called the Space Freeport Study. Among its other objectives, this study is designed to investigate and evaluate the feasibility and desirability of one international "Earthport" launch center at the equator rather than several scattered private sites. Such a location would take advantage both of the faster Earth spin at the Equator and of the savings in fuel costs that would result from an equatorial rather than a middle-latitude launching. Earthport would be a politically neutral international site whose comparatively inexpensive launching services would be available to all for peaceful purposes, regardless of ideologies or the vagaries of international politics.

Whether the direct participation in space exploration and habitation by private industry, grass-roots organizations, and individual citizens occurs or not, the space transportation systems, space construction bases, and large permanent space communities "are inevitable—these things will happen." So asserts John Yardley.

CHAPTER 2

The Space Debate

To migrate or not to migrate—*from Earth*—is the question for this age.

The launching of Sputnik and the moon landing shocked and stimulated the human imagination to reflect anew on humankind's destiny. The time for initial reflection is nearly complete, and the early elements of a serious Space Debate are beginning to appear.

Space Migration—a Page from History?

Dr. Roy Bryce-Laporte, Director of Research, Immigration and Ethnic Studies, of the Smithsonian Institution, is intrigued by the unusual dynamics involved in the popular surge toward space migration. He has studied extensively the history of human migration on Earth and has come to some instructive conclusions for the Space Debate: "People form linkages to move from less to more satisfaction. *Their motivations range from escape to exploration. [italics added]*"

The pioneer family. *Courtesy of International Communications Agency*

The majority of people have migrated when their food, firewood, or some other resource became scarce. Others have moved on to escape an intolerable war or religious oppression. The truly fortunate ones, Dr. Bryce-Laporte claims, are those who "have migrated to explore the unknown, driven by a curiosity nurtured from a surplus situation." He believes that the most peaceful migrations were those of groups escaping religious persecution by moving to new and uncontested lands, or to lands that were not already inhabited.

The religious underpinning itself was not the decisive factor in the peacefulness of the migration. Rather, the religion allowed such groups to have a shared purpose that transcended the individual; the rules of the migration game were well understood, allowing for a high level of trust.

During the present day, another kind of peaceful migration occurs among certain nomadic groups that move without claiming turf, having only a passing interest in it. Tribes in India follow the growth pattern of certain edible berries without claiming land along the way; they are quite peaceful. Of course, the worst forms of migration were—and are—those

Pioneers pole a boat down the Tennessee River. *Courtesy of International Communications Agency*

motivated by economic and military conquest: e.g., the Spanish conquistadores, the invasion of North America by Western Europeans, and contemporary forms of warfare based upon political and economic imperialism. In this context, the purpose is to invade territory that is already settled.

What Makes Space Migrants Different?

Space migration involves an unusual set of factors, giving birth to what may be a unique set of social dynamics. The linkages now being formed by future space migrants cut across national, religious, political, and economic boundaries. While previous migrations were generally homogeneous in one or more of these categories, those now wishing to leave the earth form an extraordinarily eclectic group.

Even the motivations to leave are wide ranging and diverse. Some feel that earth's resources are dwindling, and an expanding population makes the eventual migration a necessity for survival. Others feel the endless human imagination and ingenuity will always find ways for the earth to provide sufficient resources; these people simply want the stimulation of new challenges. Hence, both *escape* and *exploration* motivate future space migrants. The migration stands a good chance of being peaceful since migrants would be moving essentially to unclaimed turf. (At least, we're not aware of any extraterrestrial claims.) The issues left to resolve are the establishment of a common transcendant purpose, and a set of rules for the migration game that will instill trust. But the fact remains that Dr. Bryce-Laporte's criteria are well on their way to being met.

A Moral Issue?

Environmentalists are divided on the issue of space travel. On the one hand, says one group, it looks like the same old Industrial Age disregard for nature extending itself to the heavens. Spokesmen as diverse as Lewis Mumford, Dennis Meadows, E. F. Schumacher, and Ken Kesey are representative of the belief that the industrial mentality is inherently destructive of both the environment and of people. It is macho, prideful, irreverent, competitive, and warlike. They believe space migration merely will continue on a grander scale the worst mistakes

Pioneers formed wagon trains as safeguards against hostile natives and the forces of nature. If "star wars" become an early factor in space migration, space shuttle trains wending their ways to planets through astrophysical valleys and plains of our solar system may not be uncommon sights. *Courtesy of International Communications Agency*

of industrial civilization and will make even the cosmos accessible to the pathological sicknesses of human culture.

On the other hand, other environmentalists say space migration just might be the necessary move to stimulate world cooperation, save the environment, and embark on the next natural step in evolution. Margaret Mead, Stewart Brand, Paul Ehrlich, and California's Governor Jerry Brown, among others, have spelled it out in some detail. They believe that space migration will eventually result in large-scale industrial production being moved off the earth and into free space where the destructive byproducts can be more easily contained. This would allow the Earth of the future to become a new "Garden of Eden"—the natural environment and a highly aesthetic, miniaturized technology blending effectively and harmoniously.

They draw a parallel between space habitats and island cultures. The latter, they point out, tend to be more passive, if not peace-loving, than contiguous continental civilizations. Space habitats, as a form of island culture, are likely to exhibit the finer values of trust, cooperation, and mutual caring just as island cultures tend to here on earth.

The antispace environmentalists see the technological development of the last two hundred years as marking an unfortunate turn in the movement of human history. They fear that the synthetic environment of space communities will mean a further exaggeration of what they see as today's antihuman trend. The necessary emphasis on tools and technology will, they believe, serve only to dehumanize and despiritualize people even more. But such a sweeping condemnation of the technological advances of the last two hundred years doesn't see the forest for the trees; it rejects out of hand all the good that has been achieved along with the bad.

Basic to the antitechnological view—and one of the first premises of the entire Space Debate, whether people realize it or not—is the ages-old question: "Is human nature essentially good or bad?"

If humans are a cancerous growth in natural evolution, then it is noble to be against space migration. But according to the lessons of various religions and the apparent laws of natural evolution, humans have every right to be here. We may have chosen for the moment to foul our own nest, true; but our mobility and awareness also give us the capacity to clean it. In fact, according to the National Wildlife's EQ Index, the national level of sulfur dioxide has dropped over 25 percent since 1970.

Humans consistently make mistakes and self-correct; we save ourselves. If we were essentially bad, we could not— would not—do that.

The Salvation of Mother Earth

Any system is better understood by stepping outside it. Perhaps we must leave the Earth to help it. Seeing our birthplace from space communities might well provide a healing new perspective. Earthly self-correction might even be speeded up by space migration. Or, as Margaret Mead put it:

"We are at a point in history where a proper attention to space . . . may be absolutely crucial in bringing this world together. . . . The Pope blessed exploration of outer space as an extension of the human spirit. . . . The first pictures brought back from the moon changed the perception of earth itself."

This new perception is already helping humanity correct some of its mistakes.

There is also every reason to believe that space migrants, or Spacekind, will develop a deeper sense of empathy for all of Earth than has Earthkind itself. This is suggested in no small way by the shifts in attitude seen in most of the astronauts— from personal to general, from national to global, from individual to universal points of view.

Another criticism finds space migration nothing more than a new version of rats leaving a sinking ship. Terribly pessimistic, it assumes that Spaceship Earth is sinking.

The "population bomb" and overwhelming international weaponry rightfully justified some of our daily pessimism about the long-range future of humankind. But things are moving in a direction to make such pessimism excessive. With India's first food surplus, the Strategic Arms Limitation Talks (SALT), worldwide communication, and the world population beginning to stabilize—all examples of self-correction—it seems unlikely that Earth will sink. Earth will abide; we really need only concentrate on the basic assumption that Homo sapiens will survive.

The basic issue at the heart of the Space Debate is not so much whether humankind will survive; it is more whether humankind will survive in a meaningful way. It should come as no surprise, then, that the United States and the U.S.S.R., each from its own perspective, are conducting space activities with the primary objective of creating solutions to terrestrial problems.

Earth care, not escapism, is the first mission of space migration.

Earning the Right to Passage

The most common criticism of space migration is: "Why go into space when there are so many unsolved problems here on Earth? When we clean up the Earth, then we will have earned the right to go into space."

The "Earth first" view essentially takes issue with timing; it is not questioned that we probably should and will migrate eventually, even if only in the far future—but not until we have matured as a species. Not until we have eliminated war, satisfied all basic human needs, learned to live in harmony with nature, and have developed into a spiritually peaceful civilization. There is no reason to believe, it is argued, that life will be more "civilized" among the stars than it is on conflict-ridden Earth until basic human cooperation and the peaceful resolution of major problems are effectively achieved. One need only look to near space to see one thousand military payloads, the argument continues. It is not at all clear that "We go in peace for all mankind." Why should we spread our dirty laundry around the galaxy? We should not migrate until our own backyard is in order. Space migration, in short, becomes the reward for cleaning up the human terrestrial act.

The "Earth first" attitude seems unjustifiably moralistic, romantic, and simplistic. Problems don't end; they just change. As one set of problems is solved, there are always new ones to confront. The argument also assumes that space migration will not help Earth. We have already seen how it very well might.

If the need for the right to passage to the stars applies, but events keep moving as they have toward space migration, future historians may have to tell us what the right *was*.

Paolo Soleri, the Italian synthesizer of architecture and ecology, believes the right will be found through the achievement of a worldwide consensus. Those who wish to go must create meaningful dialogues in all fields and varying cultures to determine the most effective, healthy, and civilized steps to the stars. Such dialogue and exchange will come through the newspapers, television, and books like this one.

The clearer, more open, and more honest the exchange, the faster will come the day for migration. If the dialogues evolve into a genuine consensus that has integrity, then the right to passage will have been achieved. Paolo Soleri has little doubt that this will happen. His conviction of the rightness of man's migration to the stars is so great, he even goes so far as to say that to hold it back is to work against the Divine Logic of life!

Ellis Island served as the gateway to the United States for some 12 million immigrants, perhaps much in the same manner as Earth-orbiting space facilities will serve as way stations for migrants to the outer planets of the solar system. *Courtesy of International Communications Agency*

Money in the Wrong Places?

Another objection to space migration is cost: "Why spend all that money in space when we have so many social problems right here on Earth that money can help correct?" There are some surprisingly direct rebuttals to that objection.

The money spent here on Earth for space migration causes a healthy economic ripple effect in all areas of life. It creates jobs, helps reduce inflation, adds to the favorable balance of trade, and returns more money to the economy than was originally paid out (see Chapter 6).

Furthermore, according to a study carried out by the Chase Associates, fourteen dollars are returned to the Gross National Product of the United States over a twelve-year period for every one dollar spent by NASA research and development. Even half that figure would justify the investment.

It is already obvious that rather than wasting money, the space program directly and indirectly generates money; it does so at such a rate as to be a net profit to the government within nine years. This return on investment can go directly into social and environmental programs.

So it would seem that the cost argument against space migration should be turned around to read: "Why not spend all that money on space migration to generate the necessary surplus wealth to pay for social programs right here on Earth?"

The "Earth Sitters"

"I love Earth. I was born here, and it is here in this finite world that we will come to terms with being human." This is the key perception of so-called Earth sitters, who are viewed with disdain by supporters of space migration. To run to space is to run from ourselves, they believe.

The attitude is reminiscent of the pre-Copernican view that the Earth is the center of the universe. It is a perfectly respectable point of view. But when it becomes self-righteous, when its proponents say that no one else can leave either, it becomes unacceptably totalitarian.

Curiosity to go beyond the usual and the known cannot be suppressed forever. It is not difficult to imagine space enthusiasts, hardened by years of social ostracism, becoming militant.

Eventually, those who believe Earth is the only rightful place for humans must come to accept that those who want to go, should be allowed to go. It's only a matter of time.

Are Space Migrants a "Different Breed"?

By letting space migration happen, yet another argument proceeds, a new problem potentially arises. Earthkind and Spacekind might—indeed, probably should, considering all their differences—evolve as two distinctly separate cultures.

Space migrants may want independence from Earth faster than their earthly colonial investors are willing to allow. This could lead to small colonial wars and eventually even to a total war between Earthkind and Spacekind. Even well-disciplined astronauts, it is pointed out, went "on strike" during a Skylab mission when there was a difference of professional opinion.

But before this issue of the total distinction between Earthkind and Spacekind looms too large, it is essential to recognize that the still Earth-bound human species does not see itself as unified now. Instead, humanity sees itself as multi-colored; or American and Soviet; or man and woman; or Arab and Israeli. . . . We are even divided by language and imaginary dotted lines on maps. The very existence of a military presence in Earth orbit proves we do not see ourselves as one family. So even should space migration result in some separation, it will not divide a perceived unit, because such unity does not now exist.

A little imaginative thinking helps reduce the "different breed" argument to absurdity. As generations of Earthkind and Spacekind evolve, a new interest in mutual development may evolve. Earthkind and Spacekind may create a reunion beyond the present scope of our understanding. The step from human to humankind may well be accomplished by the intermediate step of evolving Earthkind and Spacekind. Hence, it is not clear that the creation of two distinct cultures—perhaps even species—will result in Star Wars. Indeed, space migration may

How temporary will the first space community be? Is there a factual or symbolic relationship between the first day at Jamestown, Virginia, and the first manned lunar landing? Can we anticipate a flood of migrants into space similar to that which followed the landing at Jamestown? *Augustus Lynch Mason, 1884, Courtesy of International Communications Agency*

well be the evolutionary phenomenon to bring us together in the long run.

Some Surprising Support

Space migration certainly cannot be called an evil act. In fact, a growing number of people believe that space migration fulfills various religious predictions and suggestions of the need for earthly transcendence. Hindu, Islamic, Buddhist, and Judeo-Christian prophecies elevating humankind to higher dimensions and other worlds would quite literally be fulfilled by space migration. Whether that was the original intent of the earliest storytellers may never be known. But in the meantime, the recent and rapid growth of fundamentalist religions may produce some surprising enthusiasts for space if belief in that fulfillment continues to grow in popularity.

Planetary Pirates and Star Warriors

Far from leading to an ascension to heaven, Paul Csonka, a physicist at the University of Oregon, believes that space migration will lead to tyranny and ultimately to wars among Spacekind, as well as between Spacekind and Earthkind. He is not alone in this view, but he is a leading spokesman for it. Csonka envisions a future with thousands, perhaps millions, of former colonies—and subsequently independent communities—roaming in the solar system like pirates on the high seas.

This fear for the future ignores the present emerging international concept of the pluralistic uses of common resources. It is a great part of the realistic negotiations now going on that deal with the high seas and various ocean resources—albeit still within the framework of nation-state competition for physical wealth and power. If we are capable of such cooperation here on earth, why not assume we are equally capable of it in the solar system?

The Motivation of Third-Kind Encounters

A more fanciful issue of the Space Debate encompasses the question of whether extraterrestrial intelligence exists. Astronomer Carl Sagan has argued for years that the really big issue in space lies in making contact with other intelligences. Since our space ships go so slowly relative to the vast expanse of the universe, he feels it is rather pointless to try seeking out alien intelligences much as the ancient explorers sought out hidden peoples. Instead, we should send radio signals at the speed of light, hoping someday, possibly thousands of years from now, to get a response.

This view assumes that an advanced technological civilization will be peaceful. But there is no evidence to believe that technologically superior extraterrestrial cultures will have peaceful intentions in contacting us. We simply do not know. By going out to meet others, our science and technology will advance, perhaps giving us some military and cultural parity. In this respect, one argument put forth is that by traveling out of our solar system, we will have earned the right to passage for the sake of extraterrestrial contact.

The compelling urge to stretch Earth's bonds. *Painting "Icarus" by Ascian; photographer, Carolyn Amundson*

Resistance to the New

Dr. Isaac Asimov adds a different perspective to the space debate. He studied a variety of social changes induced by technological advances and found there was always initial resistance. While a student in college, Asimov predicted that a resistance would develop to space travel just as it has to all major technological advances. Dr. Asimov believes this Space Debate is natural and to be expected, but that eventually human curiosity will prevail.

And So . . .

The Space Debate goes on. Albert Einstein once said that when two people disagree, it is probably because they are both wrong. It may also mean that both are equally right. And so, those who want to go will get it together; those who don't, won't. The space shuttle *Enterprise* has flown, orders for the future are in, and the research and development budgets are being spent.

The future—even without a date—seems clear: Space will be populated.

CHAPTER 3

Washington
and Space Migration

Space Is "In" . . .

By the 1969 Apollo 11 moon landing, the antiscience, antitechnology, and antispace attitude was finely honed among the disillusioned American public of the sixties and early seventies. But the attitude toward space exploration began a turnaround as early as 1976 with the Viking search for extraterrestrial life on Mars, the opening of the Smithsonian Institution's extraordinarily successful Air and Space Museum that draws as many as 88,000 visitors a day—and the 500,000 letters asking President Ford to name the space shuttle the *Enterprise.*

The following year *Star Wars* went to the top of all popularity charts, making movie history, and the space shuttle flew free, shaking the media into renewed interest in the future of the space program. Since then, a President has been elected who believes he saw a UFO, *Close Encounters of the Third Kind* has been a box office smash, and a Soviet nuclear-powered satellite crashed within the borders of Canada, sharply bringing the military uses of space to international attention. Even television ads for cars, jeans, and food have featured the future-of-space theme.

... and NASA Keeps Stumbling!

In early 1978 the House Committee on Science and Technology, chaired by Congressman Olin Teague, and the Senate Commerce Subcommittee on Science, Technology, and Space, chaired by Senator Adlai Stevenson III, decided to hold their own "look-see" at the future of space. Both Senator Adlai Stevenson and Congressman Olin Teague were critical of Dr. Frank Press, President Carter's science advisor, and Dr. Robert Frosch, the Administrator for NASA, for their general lack of vision and conservative space proposals. Obviously there is a new mood regarding the movement into and the use of space; the antitechnology of the sixties seems to be over at last. The growing interest in space that the new mood reflects is catching the serious attention of growing numbers of Congressmen.

View from the White House

President Jimmy Carter has not yet reflected the rapidly mounting public interest in space in any of his Administration's policies. That may change; it may have to.

Despite his lack of space policy, President Carter is taking the first steps necessary for the future development of space to maintain a coherent direction. On March 28, 1978, he asked the National Security Council to carry out a classified review, called Presidential Review Memorandum-23, to determine American long-term space goals. The result should lead to a more coherent national space policy and a program for the peaceful uses of space. PRM-23 is a complex, far-reaching study. It must determine—and resolve—the conflicting interests among NASA, the Pentagon, private industry, the Departments of the Interior, Commerce, Agriculture, State, and others, all with direct and different interests in the growing uses of space.

Dr. Zbigniew Brzezinski, head of the National Security Council, is no stranger to the issues of space. According to a

January 17, 1977, article in *The Village Voice*, he believes
"Space exploration is probably the most dramatic example of
the human adventure made possible by science, but currently it
is almost entirely monopolized on a competitive basis by the
United States and the Soviet Union. The pooling of Western
European, Japanese, and American resources for a specific joint
understanding would do much to accelerate international
cooperation."

It looks as though America could have a space program led
by the President of the United States for the first time since the
late President John F. Kennedy decided Americans should land
on the moon. Unfortunately, however, the new Administration
has inherited a complex set of unresolved policy issues. Without clear leadership from previous Administrations, the Departments of Defense, Commerce, the Interior, Agriculture,
NASA, and aerospace industries have each developed their
own preferred futures for space without adequate coordination
with the others. The result has been a fragmented approach to
space exploration. Granted, the space shuttle nicely ties many
of these interests together, but the shuttle is a vehicle in a space
transportation system; it is an implementing tool, not a goal in
itself.

Specifically, the objective of Presidential Review
Memorandum-23 is to weigh the significance and possible
implementation of: (1) political goals; (2) economic consequences; (3) space warfare; (4) national security; (5) fiscal
realities; (6) international cooperation; (7) benefits to humanity; and a classified eighth issue.

Could space migration be seen as the direction for the
space program that would resolve the various interests? No,
says the NSC. Even though the President has called for a long-
term look into the future, the building of space communities is
not included. "Government policy hasn't gotten that far," says
one White House source. "The realities of five-year plans and
zero-based budgeting put it out of our considerations. Such
issues as space communities will go to Frank Press, the President's science advisor, after the National Security Council
finishes its review."

Although President Carter promised new life for science in

his Administration, there is still a long way to go. The percent of Federal research and development money in the controllable part of the budget dropped from 14.7 percent in 1968 to 13.8 percent in 1978.

Frank Press's civilian space review, developed while the Department of Defense and the National Security Council were formulating their more immediate perspectives, did not seriously study space migration as national policy. "You're talking about a $100 billion program. . . . You won't see it in this century," predicted Press. However, the scalding Press's conclusion received from both the Senate and the House of Representatives may have some effect on the results eventually presented to President Carter for his consideration.

"The President is open-minded about space," says Dr. Press, head of Carter's Office of Science and Technology Policy. "The full utilization of the capacity of the space shuttle is our first objective." But the use of microwaved energy from space is not part of the "full utilization of the capacity of the space shuttle." Dr. Press seems to believe that the public will think that is too remote, too fanciful to be considered a serious national goal. But that was before H.R. 12505 Solar Power Satellite Bill passed the House of Representatives by a 267 to 96 margin. Press understands it is technically feasible and possibly an economic godsend to the United States. But, like the National Security Council, the Office of Science and Technology Policy is up to its ears in legal and policy questions about Earth resources identification, navigation, and communication satellites. Interestingly enough, the President's science advisor refused to speculate about the future of space development—even though many foreign diplomatic representatives in Washington, D.C., welcomed the opportunity to discuss it.

In the meantime, President Carter is quite concerned about the growing possibility of actively hostile powers in space. Recently he has suggested to the Soviet Union that each side "forego the opportunity to arm satellite bodies, and . . . destroy observation satellites."

"If we allow the military to become too dependent on space, the Soviets could aggressively follow. We're quite concerned about that," stressed a White House official. "The President wants a comprehensive space pacification program."

The alternative to a space pacification program may be a space arms race; no nation can view that possibility with comfort.

Cooperation on an international level, involving both the scientific and financial communities worldwide, that would result in energy production, the development of pharmaceuticals, the manufacture of ball bearings, and the like, all being put into space seems like the most rational way to pacify space. Such an endeavor would serve to further political goals, bring about gains in human righfs, generate wealth while reducing both unemployment and inflation, reduce the threat of space war, and provide benefits for all humanity. It is almost beyond comprehension why Dr. Brzezinski's and Dr. Press's space studies are not taking the possibility of such a program into account. Members of Congress are equally baffled, and they are trying to do something about it by passing the solar power satellite bill.

Congress Considers Its Own Space Policy

Congressman Don Fuqua wants a long-range space goal. As Chairman of the House Subcommittee on Space Science and Applications and as the expected 1979 Chairman of the full Committee on Science and Technology, he found himself dissatisfied with NASA's "Outlook for Space" study and held his own hearings in 1975 to see what the future held. After listening to such space pioneers as Arthur C. Clarke, Krafft Ehricke, and Gerard O'Neill, Congressman Fuqua believes the best candidate for our next major push in space is the solar power satellite that can send energy to Earth via microwaves. "President Carter could make a commitment similar to President Kennedy's for the moon landing," Congressman Fuqua has said. "[Such a project] is peaceful, the third world would benefit, we need it, and it will help international cooperation."

To that end, he met with Bert Lance, then Director of the Office of Management and Budget (OMB), and James Schlesinger, Secretary of the Department of Energy. Con-

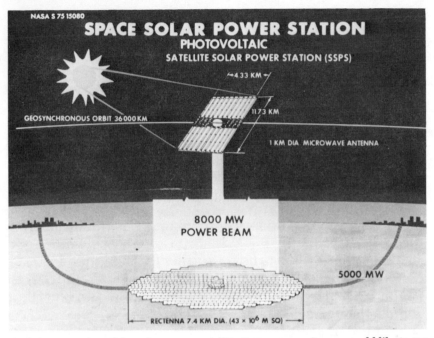

Artist's conception of the solar power satellite Congressman Fuqua would like to see as a United States national goal. It is more likely that 10,000 megawatt satellites will be used than the 5,000 megawatt one above, suitable for Houston, Texas. *Courtesy of NASA*

gressman Jim Wright, the Democratic House leader, also made the same effort. But OMB and the energy advisors called it a middle-priority item. Undaunted, Fuqua turned back to Congress to get the ball rolling. Along with congressmen Larry Winn, John Wydler, and Olin Teague, he intended to enter a resolution that would show Congressional support. The Office of Technology Assessment was to conduct a full study of the possibility of putting the first solar power satellite into orbit—a national goal to be accomplished by 1992, the five-hundredth anniversary of Christopher Columbus's discovery of the New World.

Dr. Frosch, Administrator of NASA, protested vigorously to Congressman Teague, Chairman of the House Science and Technology Committee, that this was not possible and that no such resolution should be entered into the record. Congressman Teague, at the strong insistence of Barbara Hubbard of The Committee for the Future, a citizen's space group, did change the goal date to the year 2000. But Congressman Teague proceeded to enter Concurrent Resolution H.R. 451 despite all objections. Simultaneously, Members of Congress Barbara Mikulski, Lindy Boggs, and David Stockman entered the same text as House Concurrent Resolution H.R. 447.

In January 1978, Congressmen Teague and Fuqua held hearings to develop the future options in space. Frank Press and Robert Frosch reviewed many of the past achievements of the space program and described possible future institutional assignments involving the cooperative use of satellites. Members of the House Committee on Science and Technology were disappointed at what they heard. They were slightly angered by the Administration representatives' "lack of vision."

A corner had clearly been turned. The committee wanted a long-range space goal on the same scale as the Kennedy moon-landing goal. But the President's science advisor and the NASA administrator had nothing to offer which remotely resembled a bold future in space. If President Carter would not come out with an historic space goal, then the committee would do so by backing H.R. 451 and 447.

Boldness? Brave New World? . . . or Wishful Thinking?

To add teeth to the committee's feelings, H.R. 10601 was entered January 30, 1978, just after these hearings, by Congressman Ronnie Flippo (for himself and Messrs. Teague, Fuqua, Walter Flowers, Mike McCormack, Larry Winn, Roland Gammage, Louis Frey, Dan Glickman, Jim Lloyd, James Blanchard, and Dale Milford). This bill called for $25 million for a study of the feasibility of manufacturing solar power satellites that includes research, development, and demonstration. On May 2, 1978, it was reentered as H.R. 12505 by Congressman Ronnie Flippo (for himself, and Messrs. Teague, Fuqua, McCormack, Winn, and John W. Wydler) and passed.

Congressman Fuqua was confident these bills would be passed by the House Committee on Science and Technology and sent into the House, bringing about the start of Congressional debate on the Second Age of Space. "This is important," he said. "The nuclear advisors see it [space manufacturing and migration] as competition, but it fits with the President's policy of preventing nuclear proliferation." Fuqua seemed matter-of-fact about it. "With our standard of living, we cannot compete with menial work. Technology is our bag of tricks. We must stay ahead technology-wise. We have the capital to do it, and it seems like the best goal."

All things taken into account, a thousand-megawatt nuclear reactor costs $2 billion. It will require 531 of them to satisfy only our current electrical needs, according to the Atomic Industrial Forum. That's over a *trillion dollars*. The solar power satellite program is estimated at under $300 billion. "But it all takes more study. No one really knows the costs," said Fuqua. Hence the point of the Congressman's resolution. He admits the satellites would be hard to defend in time of war. "But so are Earth-based resources," he added.

Space shuttle assisting in the construction of a solar power satellite. *Courtesy of NASA*

Although quite reserved throughout this discussion, Congressman Fuqua became very animated when he spoke about human nature and the space program. "We've always pushed ahead where we've never gone before. It's the fascination of mankind to challenge the unknown. . . . With the solar power satellites will come space industrialization and human colonies." He explained that the Congress cannot dictate long-range plans, but it can make judgments and establish priorities. It's not difficult to tell that this Congressman sees space as one major key to the American future. "This is something we're doing technology-wise that was never done before."

Jolt of the JOP

The first visible sign of the new Congressional interest in space was spurred by an overwhelming turnaround on the funding of the Jupiter Orbital Probe (JOP). NASA had asked for $20.7 million for JOP in the fiscal 1978 budget, leading toward a launch date in 1982. If the planetary spacecraft were not launched at that time, there would have to be a wait of several years for another, less favorable, launch date.

Three of the four Congressional budgetary committees approved NASA's request. But House Appropriations Subcommittee Chairman Edward P. Boland refused. He felt NASA had enough with the space telescope and should get only one big-money project a year. Besides, Boland argued, "Jupiter will always be there." He wasn't fully aware of the critical launching time. Apparently he also did not realize that the loss of funding could have shut down the Jet Propulsion Laboratory (JPL), a highly specialized team of planetary experts whose close working relationships were extremely valuable for the continuity of the entire space science program.

NASA, the aerospace industry, and other space experts felt JOP was lost. But the concerted effort of Governor Jerry Brown, grass-roots space organizations, and Gene Roddenberry turned the Congressional vote around. Roddenberry, the producer of *Star Trek*, sent a telegram to Nichelle Nicols, the actress who played Lieutenant Uhura in the TV science fiction series. She happened to be speaking to a Star Trek convention in Philadel-

phia at the time. The telegram urged the fans to run to their
hotel rooms, call home, and urge all their friends to send tele-
grams and night letters to their Congressmen to vote yes for
JOP. It worked. The day before the vote, JOP was expected to
lose by 2 to 1. Instead, the bill passed by 2 to 1; the veteran
Congressional watchers for the space program were stunned.
Congress was surprised, too. Public support existed; politics
being what they are, Congress took note. H.R. 451, 447 and
12505 demonstrate clearly that Congressmen formerly shy on
the issue of space have become more bold.

The Senate

The Senate, too, has picked up some of the new space
tempo. In February of 1978, Adlai E. Stevenson, Chairman of
the Senate's Subcommittee on Science, Technology and Space,
held a symposium on the future of space. The symposium
dealt only with the immediate applications of space for global
communications and earth resource identification.

Senators Stevenson, Ernest Hollings, Donald Regal, and
others seemed primarily concerned with how to sell the space
program to the public. They apparently have forgotten that the
public has begun to react favorably to the possibilities of our
future in space. Although the Senate is more conservative than
the House on the scheduling of an expanded space program,
the Senate committee members feel it is the correct direction for
the future and that it will yield great benefits for the United
States.

Senator Stevenson pointed out that NASA space activities
contribute $200 million annually to our balance of payments
and that "it may be vitally important in the years ahead for us to
understand fully the space environment in which the earth
rotates, not only from the standpoint of exterior impacts on our
planet but also for the effect of our own activities on that
environment. This is the principal goal of our physics, astronomy, and lunar and planetary projects." He also felt that it
provided a common ground for all peoples leading to a more
peaceful future. "The fundamentals of science and technology

are identical [for] all nations and all peoples. This fact provides a common base for progress."

Dr. John Stewart directs the Senate committee's staff work on space. He pointed out that "space is a nonpartisan issue. It is not conservative or liberal. The support is across the board."

Probably the two strongest supporters of space in the Senate are Barry Goldwater, a conservative Republican, and Adlai Stevenson, a liberal Democrat. The only really antispace Senator is William Proxmire.

These are great strides forward. But they pale in comparison to the possibility of space migration. Even Dr. Robert Frosch of NASA felt that he was being put in the role of technological conservative for not pursuing space migration and manufacturing as aggressively as some would like.

Dr. Stewart explained why the Senate has not been too visionary on space: "The Senate is much better on near-term issues, since few [members] will be running in thirty years. Space industrialization and migration are thirty-year issues." He does feel that the Senate has been pro-space from the beginning. "The Shuttle has been supported right along. We prefer an evolutionary step-by-step approach. Don't expect a twenty-year goal. You get more for your money if you go slowly. The United States will still be in the forefront with increasing international cooperation." He feels the big moves in the future will be international: "The internationalization of the space program will (1) spread the cost; (2) reduce suspicion and lower military expansion; and (3) space doesn't belong to any one nation—who's in charge of geosynchronous orbit over nations?"

Like State and Defense Department officials, Dr. Stewart does not see the space treaties nullified by Soviet FOBS— Fractional Orbital Bombardment System (part satellite, part ICBM)—or hunter-killer satellites. "We've been reasonably successful at keeping the military out of space," he feels. A surprising comment, considering that the majority of hardware in space and most space funding is for military purposes!

Stewart does agree that a new public interest in the peaceful uses of space is gaining momentum. He points to the JOP vote and the Smithsonian Institution's National Air and

Large space telescope being placed in orbit by space shuttle. It will show the largest, most detailed view of the distant reaches of the universe by the mid-1980s. *Courtesy of NASA*

Space Museum, the most popular tourist attraction in Washington, D.C., as proof.

Governor Jerry Brown of California

Meanwhile, a new force has entered the politics of space. Among those in positions of political leadership, Governor Jerry Brown of California has one of the clearest, most philosophically thought-through perspectives on space technology's role in human evolution. "Going into space unleashes great imagination for new and positive visions of the future. As we see our true place in the universe, we will reach a unity of mankind. This leads to a more profound understanding which diminishes our provincialism."

On the eve of the first solo flight of the space shuttle *Enterprise*, Governor Brown held a symposium to celebrate California's first Space Day, declaring that "Small is beautiful on Earth, but in space, big is better." He understands that one can consistently support moves toward small-scale, decentralized earthly programs while pushing for space migration. Spacekind and Earthkind can each have its own way.

"I also think [about] the closing frontier, the closing of the West, and what that does to the psychology of people," Governor Brown goes on. "As long as there is a safety valve of unexplored frontiers, the creative, the aggressive, the exploitive urges of human beings can be channeled into long-term possibilities and benefits. But as those frontiers close down and people begin to turn in upon themselves—that jeopardizes the democratic fabric." The validity of Governor Brown's last concern is reflected in the results of Dr. John Calhoun's experiments with mice at the National Institute of Mental Health. The mice were provided with all the necessities of life—but without extra space, and with no challenge, no competitive situations, no "new frontiers." The mice population grew to a peak of 2,200, then fell at the end of three years to only forty-six sluggish, aging, asocial mice.

Governor Brown went on to say in his Space Day address: "You can't limit the mind of science and technology. . . . [Space] is an expanding asset, and through the creation of new wealth we make possible the redistribution of more wealth to those who don't have it. . . . When the day of manufacturing in space occurs and extraterrestrial materials are added into the economic equation, then the old economic rules no longer apply. It's just a question of the political will and the leadership, and the imagination, and the ability of the private sector, the federal and state governments, the universities, working together and sensing the potential."

As the first governor to put an astronaut on his staff, he is preparing California to be the leading state in space by signing the first memorandum of understanding for space applications between NASA and a state, and by setting up a space institute in the University of California system. He is also the first governor to have his state buy into a satellite—Syncom IV.

Looking beyond his own state, Jerry Brown sees space migration as a profitable way to stop the madness of the arms race. "One of the problems in turning around the arms race is jobs. An expanded space program can employ the same people." A joint U.S.-U.S.S.R. program to build solar power satellites would divert military spending, Governor Brown adds. He has discussed these ideas with delegations from the Soviet Union and China and says they have been received with enthusiasm.

Brown sees space in a spiritual context as well as understanding its political and economic benefits. "As we develop space, we will see ourselves as a single species; hence, space manifests an opportunity to bring people together as never before."

The major religions of the world have taught that all people are one in spirit. This sense of oneness, Jerry Brown points out, was consistently experienced by the astronauts. Rusty Schweickart, his astronaut assistant for science and technology, has spoken about the spiritual effects and changed awareness induced by space flight.

Governor Brown has actively lobbied in Congress for the Jupiter Orbital Probe appropriations and for H.R. 12505 to study space manufacturing and the production of solar power satellites by the Department of Energy with advice from NASA. He sees the whole political and public climate of opinion opening up to allow for a new space vision on a scale comparable to John F. Kennedy's mission to the moon. But this time he sees it in cooperation with the Soviet Union.

Politics and Space

Policy analysts in the conservative and moderate Republican ranks have begun studying future space possibilities. They see the industrialization of space as the new frontier for private free enterprise. These analysts seem to realize that space industrialization could salvage the free enterprise system and even get this country off the nuclear hook if solar power satellites can work. At the same time, President Carter is developing his space policy and Governor Brown is getting national recognition for his leadership in bringing space to the attention of the American people. Hence, the Democrats and Republicans will have space increasingly on their minds as the 1980 presidential primaries draw near.

If Governor Brown decides to run for the presidency, it is an excellent bet that space will become an issue in the 1980 presidential campaign in much the same way as his predecessor Governor Reagan brought the Panama Canal to political attention. Jerry Brown is the political wild card that could make the Washington space connection.

CHAPTER 4

Public Support Mounts

Who Gets to Go?

When the powers that be decide to make the big push into space, who gets to go?

You.

If you really want to be part of the next chapter in the evolution of life, make some noise, kick up a fuss, find out who makes the decision and how, and figure out what you've got to offer.

Freedom seekers, curious adventurers, capitalists and communists—everyone who wants to should have the chance to go.

So far, it's been the astronauts of the United States and the cosmonauts of the Soviet Union, Czechoslovakia, and Poland. Plans are being made to include European scientists in the early shuttle flights. So it seems that American, Soviet, and European scientists and engineers will get the first crack. But if Gerard O'Neill's plans and other plans that will no doubt come along get accepted, then people from every culture and every walk of life become eligible to populate the cities in space.

The more politicians feel the pressure of a space-conscious public, the better your chances are. The more the public demands space migration, the more corporations will take the financial risk of the research and development needed to satisfy the coming Spacekind market.

Gerard O'Neill believes it is possible that more human descendants will be in space than on Earth by the year 2150. Pat Gunkel, formerly of the Hudson Institute, writes in his *The Future of Space: An Encyclopedic Prospect* that he too expects more Spacekind to exist than Earthkind—by 2105, forty-five years earlier. But what finally happens will be determined by world public opinion. Politicians react and the media reflect public sentiment. Make yourself heard—voice your opinion loud and clear!

O'Neill claims that "The important thing is that in all of this time [seven years] not one major concept necessary to the design of a space colony has been disproven." The engineering is not what is slowing down space migration; it's the lack of vision. Does anybody care? Who's whipping up grass roots support?

It should be clear that NASA does not see its job as whipping up that support. The policy seems to be to keep any visionary leadership it might possess under tight control. Maybe Jimmy Carter will change all this. But in the meantime, the real spearhead of public interest lies in the work of Dr. Gerard O'Neill. He has made it possible for a small handful of pressure groups for space to have something tangible to point at.

The Committee for the Future

The most fervent and long-lived advocacy group for space is The Committee for the Future. Beginning in 1970, it wanted to find the money, buy the rockets, and launch a citizen-sponsored moon landing. The committee called the project "Harvest Moon," and set up a second corporation called the New Worlds Company to sell shares of no-par-value common

Barbara Hubbard and John Whiteside (center and left) of The Committee for the Future, looking at the kind of space hardware necessary for Harvest Moon. *Courtesy of CFF*

stock to anyone in the world. The money would buy launch vehicles and command and lunar excursion modules from the Apollo program, along with NASA services to place experimental human habitations on the moon.

Needless to say, NASA, the Department of Defense, certain international federations, and some members of Congress resented this intrusion into "their" territory. Undaunted, Barbara Hubbard and John Whiteside, the cofounders of the committee, held meetings, gave speeches, toured Europe, and even went to Moscow to drum up support. People loved the idea, but the stocks were hard to sell since there was no clear way to guarantee a financial return. On April 15, 1971, NASA said it would not support the idea. This killed all attempts to sell stocks.

Nevertheless, The Committee for the Future testified before the House Science and Astronautics Committee on March 9, 1972. As a result, Congressman Olin Teague introduced a resolution:

Courtesy of CFF

That the National Aeronautics and Space Administration be
instructed to reserve a Saturn 5 rocket, a command service
module, and a lunar module for possible use by a transna-
tional mission; and . . . That the National Aeronautics and
Space Administration shall be and it hereby is authorized
and directed to provide technical advice to the Committee on
Science and Astronautics of this Congress, and to the Committee of [sic] the Future for the purpose of preparing a feasibility study.

In July, 1972, NASA recognized The Committee for the
Future was not going to disappear. NASA felt compelled to
explain again its reluctance to become involved with "Harvest
Moon" because of the following principal factors:
   (1) The Apollo astronaut teams had been severely up-
       rooted by the cutback in Apollo missions, and the
       restructured training program would not permit quali-
       fied personnel for a lunar expedition;
   (2) The Committee for the Future was administratively too
       unstable to complete such an undertaking; and
   (3) NASA's programs had moved in a different direction.

The project died, but a tentative agreement was reached with NASA officials to the effect that the most meaningful direct-citizen support of space would be the sponsoring of an additional space station for the Skylab program called Mankind I. The committee rewrote Congressman Teague's resolution, reflecting this new direction. But the agreement never really bloomed into a cooperative effort. The Committee for the Future was still angry with NASA, and NASA didn't want help from such an independently minded group.

Although Harvest Moon and Mankind I failed, they were not a total loss. The committee began a dialogue that is beginning to bear fruit. They convinced many leaders in public and private life that one day humans will live in space, that space living will force international cooperation, that the average person should feel some direct relationship to future space programs, and that life in space should be spoken about in philosophical and spiritual terms as well as in engineering jargon.

The second phase of the committee's activity was the development of the now well-known Syncon (for "synergistic convergence").

During their first two years of travel and talk, the members continually heard people say, "Sure—I'm in favor of international space cooperation. But the other guys won't go for it." John Whiteside got so tired of all the "Yes—but" that he decided to get representatives of as many of the different groups as possible together for a solid face-to-face exchange. It was about time, he thought, for everyone to find out just what objections to space exploration there were.

The result was Syncon, a new kind of conference. Several hundred highly diverse people got together over a period of several days to discuss and examine the various future space options. The crowd was divided into thirteen small groups, each representing major factors in society—government, technology, production, social needs, and the like. Each group reported on its goals, needs, and resources, and then merged with another group to develop a joint report. On the final day, everyone assembled and pieced together a general vision of the future, based on these reports, that included specific items for action. The entire conference was videotaped.

Syncon in Los Angeles, California, 1972. *Courtesy of CFF*

The Committee for the Future believed that once people could gain some kind of unified perspective on the future, space migration would be understood as the logical next step. It was truly grass roots education about the future. Since 1972, over twenty-six Syncons have been held across the United States and in Jamaica. Five of the Syncons were televised live with telephone call-ins for home viewer input.

To extend the educational outreach, in late 1973 the committee opened the New Worlds Training and Education Center for future-oriented activists. Students were taught how to set up a Syncon, how to manage its television coverage live or taped, and supplied with general information about future possibilities.

The third and current phase of the committee's work consists of two complementary activities: theater and politics.

Their first theatrical production, called *Previews of Coming Attractions*, is a computerized multimedia presentation narrated by Barbara Hubbard that attempts to "dramatize the history of our future and to envision and experience ourselves at the next phase of evolution . . . as Universal Humanity." The first performance was in Los Angeles in February 1978.

In the political area, the committee organized the first Congressional Seminar on Capitol Hill on "The High Frontier: Human Settlements in Space" in October 1977. It was co-sponsored by Olin Teague, Barbara Mikulski, and Edward Pattison. Subsequently, Olin Teague introduced House Concurrent Resolution 451 (with companion resolutions introduced by Mikulski, Mrs. Lindy Boggs, David Stockman, and Robert Roe).

In sweeping poetic language, the Resolution introduced a futurist philosophy that could become a "Magna Carta for the Future":

> This tiny Earth is not humanity's prison, is not a closed and dwindling resource, but is in fact only part of a vast, expanding system rich in extraterrestrial opportunities as yet far beyond our comprehension, a "high frontier" which irresistibly beckons and challenges the American genius.

Syncon in Washington, D.C., 1973. *Courtesy of CFF*

Scene from The Committee for the Future's "Theatre for the Future: Preview of Coming Attractions." *Courtesy of CFF.*

Specifically, the Resolution calls for the Office of Technology Assessment to "organize and manage a thorough study and analysis to determine the feasibility, potential consequences, advantages and disadvantages of developing as a national goal for the year 2000 the first manned structures in space for the conversion of solar energy and other extraterrestrial resources

to the peaceable and practical use of human beings every-
where.''

The Committee for the Future hopes the Resolution will
generate a national mandate by the 1980 presidential and congressional primaries, with candidates in both parties responding to the issue of growth-no growth.

Barbara Hubbard raised the issue in her testimony before
the House Committee on Science and Technology Hearings on
Future Space Programs on January 25, 1978:

> Now, in our time, there has emerged the greatest issue of
> freedom ever faced consciously—the freedom to develop our
> full capacities as a species by opening the High Frontier—an
> unlimited, nonterrestrial resource base for humanity . . . or
> [to accept] restriction, adaptation to limits, increased control
> of our lives.

The goal, as Barbara Hubbard sees it, is the opening of the
high frontier—the establishment of the first productive nonter-
restrial base by the year 2000, to operate for the benefit of all
people through the use of resources beyond the biosphere of
Earth.

Such a goal marks a quantum step in history and in evolution. It is also becoming a realistic possibility as more and more
pressure is placed on the planetary limits of Earth. The Committee for the Future can be reached at 2325 Porter Street, N.W.,
Washington, D.C. 20008.

National Space Institute Is Established

The National Space Institute is a membership organization
formed in 1973 that wants to work with the nation at large. The
members view space as a tool to help solve earth problems
rather than as an end in itself. NSI wants to inform the public of
the positive impact of space on energy and food production,
the economy, and on the general growth in knowledge. NSI
wants to become the conduit for public opinion to NASA and
the aerospace community. The members write well-placed articles for national magazines and publish a slick monthly newsletter. Although it was started by Werner von Braun and Tom

Hugh Downs, NSI board member, moderates NBC-TV show *Not For Women Only* with panel (left to right) Dr. Gerard K. O'Neill, Princeton University; Charles Hewitt, Executive Director for NSI; and Dr. Peter Glazer, Arthur D. Little and Co. *Courtesy of NSI*

Turner of Fairchild Industries and heavily financed by the private sector of the aerospace community, NSI is now receiving its support from the general public. They have added such nonaerospace personalities as John Denver, Jacques Cousteau, and Hugh Downs to their board of governors.

NSI goes to various sections of the country and holds meetings with regional leaders in agriculture, health, and economics. Their purpose is to persuade these influential people to become spokesmen in their own fields for the expansion of space activities. NSI representatives explain what space research has contributed, and the leaders are asked to speculate about how space research might contribute even more in the future. Once communication is established between NSI and the nonspace people, the latter are asked to help influence attitudes toward space in their communities and in their fields at large.

Such public awareness programs are focused on a given geographical area for three to four weeks. Mayors and governors are asked to declare this period "Future Technology Time." Speaking tours, essay contests, and special demonstrations of satellite communications are set up. This intensive

campaign ends up with "Involvement Day," featuring speakers, entertainment, news coverage, and exhibits for the whole
family.

Over the next several years NSI intends to maintain high
visibility by putting noted leaders on TV and radio talk shows
and producing announcements for the media, as well as at-
tempting some space documentaries. The National Space Institute can be reached at 1911 North Fort Myer Drive, Arlington,
Virginia 22209.

The "Trekkies" Become a Force

Neither The Committee for the Future nor the National
Space Institute worked with or tried to arouse true grass roots
support for the space constituency, for they didn't work with
the *Star Trek* fans!

The TV science-fiction series, laid 200 years in the future,
attracted one of the most dedicated followings in the history of

John Denver, top country-western singer, and the late Dr. Wernher von Braun discuss
space shuttle applications at NSI conference. *Courtesy of NSI*

A future astronaut hopeful explains the space shuttle to Lou Villegas, NSI, and George Mueller, International Aeronautical Federation, at a large model-rocket exhibition and demonstration. *Courtesy of NSI*

television. There are approximately *25 million* "Trekkies" in the United States. That's bigger than the farm vote. They span all occupations and reflect all kinds of values. It's truly an eclectic group, made up of hundreds of fan clubs, many with their own magazines or "fanzines," and newsletters. There is no central organization or headquarters. No one had successfully mobilized this sleeping giant. That is, not until a group appeared that called itself "Enterprise Frontiers." Its members decided it was time to get Star Trekdom to unite fantasy with engineering reality.

One of its members, noted science writer Richard Hoagland, called Jerry Glenn at Future Options Room to help advise the new group on how to make such a transition. As luck would have it, he was in conversation with both Rockwell International, builder of the space shuttle, and William Gorog, then-President Ford's Deputy Economic Advisor. William Clark, a business associate, and Glenn had lunch with Gorog to talk him into getting the President to give the address at the christening of the space shuttle. Gorog suggested the name "Free Enterprise," since the space ship was a triumphant expression of the free enterprise system at work. Glenn laughed and said, "Well, at least you'll get the Star Trek vote." Little did he know what he had said. That same evening Glenn met the Enterprise Frontiers group for the first time.

The group decided to try its first test of strength—a massive write-in campaign to the President suggesting the name "Enterprise" for the first space shuttle. Could the group get the cooperation of the Trekkies across the country? Could enough letters be generated to get the President of the United States to agree? Enterprise Frontiers went to work. They wrote to Star Trek clubs, set up petition booths at Star Trek conventions, and helped Glenn prepare the following brief sheet for Gorog and the President in June 1976:

THE STAR TREK COMMUNITY
Briefing sheet 6/76

The twenty-five million American *Star Trek* fans constitute a space constituency and an untapped political movement for the future. Its ten-year-old behavior pattern is not that of a fad, but of an embryonic pressure group unaware of its potential influence.

1. Five hundred thousand wrote in to NBC to save the show the first year it was threatened, and one million the second year.

2. Two hundred fifty-nine stations show its ten-year-old reruns, including forty-seven foreign stations, giving it more viewers today than ever before.

3. Currently Star Trek on Channel 20 in Washington, D.C., outdraws NBC National News, and outdrew at one point CBS National News in Los Angeles.

4. Over eight million Star Trek paperback books have been sold, plus two million sales of the bestseller Star Trek *Technical Manual* (listed at $6.95 a copy).

5. Star Trek sales have jumped 400 percent since 1971, and are now over a billion-dollar annual industry.

6. Conventions are skyrocketing. New York City alone had three this year with over 100,000 in attendance.

September 17, President Ford intends to christen the first space shuttle the *Enterprise*. The naming is not only a reference to the free enterprise system but is a direct recognition of the Star Trek community.

Meanwhile, Star Trek conventions across the country began chewing over the issue—should they get involved in political pressure? Was the future of the space program really an issue for TV fans? Could they really have any effect? Slowly at first, then rapidly, letters began pouring into the White House—at the rate of a thousand a day.

You would think that such public interest in NASA's shuttle would be welcomed by NASA. But they didn't want any help. They wanted to name the shuttle the *Constitution* on September 17, Constitution Day. One NASA official was so upset at the Star Trek write-in campaign that he told the hotel manager of the Washington Hyatt Regency to order Enterprise Frontiers to remove their sign asking people to sign the petition at the Star Trek convention. He said it was "an unauthorized write-in." Since when does a U.S. citizen need authorization to petition his government?

Meanwhile, Gorog suggested to Dr. James C. Fletcher, the Administrator of NASA, that he meet with William Clark and Jerry Glenn about naming the space shuttle *Enterprise* and for some creative ideas on the first public showing. Dr. Fletcher (now Vice President of the National Space Institute) never met with them. Finally, the President called Dr. Fletcher to the oval office September 8, 1976—ten years to the day after *Star Trek* went on national television. The President told Dr. Fletcher than the name for the shuttle was to be the *Enterprise*.

Rogers Morton, head of Ford's campaign staff, had gotten

Model of science fact NASA space shuttle and model of science fiction Star Trek USS Enterprise. *Courtesy of Richard Hoagland; Photographer, Elinore Crow*

the point: More votes would be won than lost by naming the space shuttle the *Enterprise.*

The Enterprise Frontiers won the first time at bat. It hit a home run at the Presidential level over NASA's objections.

The Star Trek fans across the country were both shocked and delighted. They couldn't believe that they had done it. But sure enough, on September 17, 1976 (Constitution Day), at Palmdale, California, a band played the *Star Trek* theme music and the shuttle was named the *Enterprise.*

> Seven members of the "Star Trek" series, including Leonard Nimoy, were at the rollout. Goldwater jokingly mentioned "a new kind of astronaut—one with pointed ears."
>
> *—Newsweek*
>
> White House sources said yesterday [September 8, 1976] that Ford agreed to follow the advice of the letter writers.
>
> *—Washington Post*
>
> Replied Fletcher (Administrator of NASA), "I can't think of a better name!"
>
> *—U.P.I.*

It is just possible, however, that one of the huge vehicle's more important contributions to the space program was reflected at the rollout by the name printed on its side in stern, sans-serif type: *Enterprise*. . . . The point is that a whole lot of people asked something of the space program—and got it. The operative difference here between Trekkies and others interested in space seems only to be that the Trekkies know that it can pay to stand up and be counted.

—*Science News*

The purists are indignant and charge that what has happened has been a press agent's coup. No doubt the charge is true. Nevertheless, tribute to all the forms of imaginative entertainment that have spurred interest in space and science this past century is long overdue.

—*The New York Times*

With success in their teeth, Star Trek fans moved next on Congress, mounting a swift telegram and night-letter campaign at the final hour to save funding for NASA's Jupiter Orbital Probe (JOP) at Star Trek producer Gene Roddenbery's direction; that worked too. Next the letter-writing machine geared up to support H.R. 12505 to have the Department of Energy study space industrialization and solar power satellites.

Students Organize Into an Influential Group—FASST

The Forum for the Advancement of Students in Science and Technology (FASST) plays a unique role. It acts as coordinator of information and activities, and offers opportunities for college students to become actively involved in the discussion of science and technology issues. It is a nonprofit educational organization with chapters on college campuses around the country, and abroad. The members testify before Congress, hold national conferences, write reports such as *Technology, the Economy and the Space Program* (1973), put out a members' newsletter, publish a quarterly newspaper, and disseminate scientific information to campus media editors through a press service.

FASST members discuss aerospace issues with Dr. John Naugal, Assistant Administrator of NASA, in the Roosevelt Room of the White House. *Courtesy of FASST*

As part of their efforts to increase student involvement, they have been encouraging NASA to develop a specific program for students' experiments to fly on the space shuttle. FASST is also working to set up student internships in NASA programs. It is putting feelers out in other national student organizations to help coordinate various student networks and possibly to hold annual national student associations conferences. Another item on FASST's future agenda is to place students on advisory boards for the American Association for the Advancement of Science (AAAS), NASA, the National Academy of Sciences, etc.

Similar to other grass roots organizations mentioned here, Alan Ladwig, President of FASST, wants to "demystify space." He sees space as part of the future work environment of students. FASST feels it is important to get students thinking and studying about space now so that they will be better prepared to be the decision makers in the year 1990, 2001, and beyond. FASST can be contacted at 2030 M Street, N.W., Washington, D.C. 20036.

The L-5 Society

One of the newest public awareness groups for space is the L-5 Society, formed in 1975 and named after Gerard O'Neill's

Ray Bradbury addressing FASST/AIAA conference on the "Search for Life in Our Solar System" at the Jet Propulsion Laboratory in 1976. *Courtesy of FASST*

L-5 space communities, described in another chapter. It is the only major space group not based in Washington, D.C. Although their 2,200 members are mostly college students, some of the leading names in aerospace are members. Their objective is to get the decisions necessary to begin space migration made as soon as possible through the adoption of O'Neill's space designs. They are perfectly willing to support other space plans; however, they feel O'Neill's are the best so far. O'Neill is more than flattered, but he keeps the group at arm's length.

The society is an activist group that descends on conferences such as the UN's Habitat or the Club of Rome's Limits to Growth conferences to talk up space migration. They see space as the monopoly of the military-industrial complex, and hence a program of the conservative rather than the liberal. They want to divorce space programs from the military-industrial complex and rekindle the Kennedy dreams. They add a more distinctly counterculture flavor to the space debate than the other groups.

The L-5 Society sees space migration as a more likely way to create a new way of life than the direct reform of earth's problems. Space is fun, and they want to go. It is this personal commitment to live in space within their lifetimes that gives the L-5 society its unique drive.

L-5's early newsletters reflected all the idealistic enthusiasm in a nonprofessional format. But they grew up surprisingly fast over their first year, and their newsletter has evolved over the past three years into one of the more sophisticated magazines among the public space groups covering advanced space research.

The society has sparked healthy cooperation among the American Astronautical Society, American Institute of Aeronautics and Astronautics, National Space Institute, FASST, The Committee for the Future by working closely with them in sponsoring conferences, coordinating Congressional action, and providing information.

FASST members and Soviet students discuss international aerospace issues with Congressman Mike McCormack. *Courtesy of FASST*

The L-5 Society plans to increase its information exchange role and is increasingly adding speakers to public forums. The L-5 Society can be contacted at 1620 North Park Avenue, Tucson, Arizona 85719.

Professional Groups Become More Activist

Traditionally professionally oriented, the American Institute of Aeronautics and Astronautics (AIAA) and the American Astronautical Society (AAS) have recently begun encouraging their membership to get out and generate grass roots support for space. With 30,000 AIAA and 3,000 AAS members, this could have significant impact on the space movement. Previously these aerospace professionals talked to each other, and performed an extremely valuable role by testifying to Congress; now their sphere is broader. The AIAA can be contacted at 1290 Avenue of the Americas, New York, New York 10019, and the AAS at 6060 Duke Street, Alexandria, Virginia.

Students tour NASA Goddard Space Flight Center at FASST conference. *Courtesy of FASST*

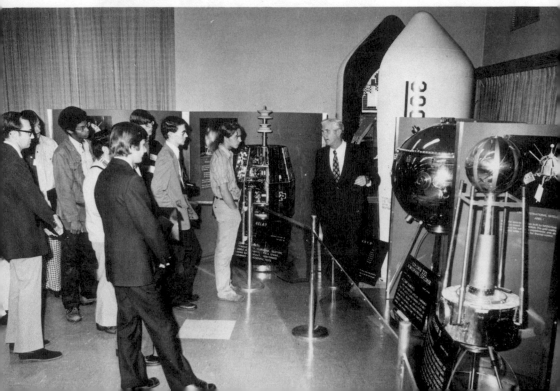

A Broad Spectrum of Support

Herman Kahn, in *The Next 200 Years*, says that space will play a much larger role in humanity's future than was ever believed; in fact, he intends to write his next book on space.

Stewart Brand of the *Whole Earth Catalog* now puts out the *Co-Evolution Quarterly*; it has strongly endorsed space migration in general and O'Neill's space communities in particular and published *Space Colonies*.

The opinions of such representatively diverse leaders, together with the various space-oriented groups, should accelerate some major turnarounds in public opinion.

Although Gerard O'Neill has been the focal point and the scientific and technological backbone of the space migration movement, he has not become the charismatic leader at the center. Nor should he be. Leadership in the space movement is the job of the grass roots support groups.

For space migrants to evolve into an international rather than remain simply an American phenomenon, the movement must involve all people from all levels, all nations, and develop a truly international point of view.

CHAPTER 5

Some Voices
from Around the World

Each step into outer space has been greeted with enthusiasm by the international community. The recent Mars landings were no exception; they made headlines around the world.

The conservative British *Daily Mail*: "It is not to find new worlds for conquest that we penetrate ever deeper into space, but to place our human life in a galactic perspective."

The business-oriented Indian *Jam-e-Jamshed*: "[the landing on Mars] has helped to put an end to talk about the futility of space exploration. The Viking's successful landing on Mars is a unique achievement of the U.S., a feat with which it has surpassed Russia in the space race. America will be remembered in human history for a long time."

The Tunisian ruling-party paper *L'Action*: "This grand victory for American science not only concerns outer space and the planets nearest to earth but offers new techniques which can revolutionize many other domains to which scientific and technological discoveries are applied."

What this begins to make plain, and what will emerge in this chapter, is that it is sheer myth to assume that only the United States and the Soviet Union care about space. The Third World cares. In fact, all countries care. The diversity of the

nations whose views are given in this chapter proves that to be
so.

Zaire, seemingly a country very remote indeed from space
flight, has German rockets being launched from its soil—and
wants to go when the price comes down, which it sees as
possible under certain circumstances explained by its ambassador. The Austrian spokesman agrees almost entirely with
Zaire's position, and adds that international cooperation in
space will help bring peace to Earth.

Surprisingly for most Westerners, not only does Indonesia
want to be involved, but space migration is also predicted in
that country's mythology—and realized already in part by their
rather complicated satellite system. The Soviet Union of course
includes space travel as part of national policy, and Iran—concerned about only Super Power involvement—offers a suggestion to avoid conflict. Colombia strongly disagrees, however,
making some cogent and sophisticated points of her own. A
Southeast Asian ambassador proposes a step-by-step plan for
world involvement—which Switzerland, as a neutral, suggests
ways to implement.

All the diplomats interviewed for this book were reluctant
to discuss the role of the multinational corporation in space
migration. Some felt they didn't know enough about the
technology of space to make recommendations. But they all felt
that America made a major mistake when it cut the space
program after the successful moon landings.

Zaire

His Excellency Ambassador Asal B. Idzumbuir of the Republic of Zaire sees expanding into outer space as inevitable.

"You cannot hold [back] technology. And who knows what you will find?" The development of space should be a cooperative venture. But since his years of experience make him doubt that effective laws regulating the use of outer space can be imposed on nations, he feels the UN should take the initiative and find what institutional arrangements would be acceptable worldwide. He disagrees with those who feel we should not go into space until all our earthly problems are solved. He pointed out that the use of satellites to locate natural resources is very important, and highly useful in helping solve earthly problems. He did express concern over spy satellites, however, and he was not optimistic that the UN could solve the military problems easily. We can only ask that nations exercise "self-restraint and not use satellites to arm" themselves militarily.

Ambassador Idzumbuir expects that the day will come when his people will visit and live in space. The cost for the moment is too great. But competition and expansion will inevitably lower it, resulting in the increased accessibility of space to all people. As a prelude to that day, and to hasten it, he feels that nations should share in the cost of space so that no one nation has to carry too much of a financial burden, and so that all nations are involved. The development of space is "for all human beings," even though "the technological leaders can get some advantage."

Austria

His Excellency Ambassador Karl Herbert Schober of Austria agrees: "One should not forget that outer space is not something where our traditional concepts of property and exclusive rights apply. I not only mean this in a legal sense—the international legal instruments, whereby all nations renounce their right 'to own property' in outer space are clear enough in that respect—but rather in a moral philosophical sense. The urge to explore, to discover the unknown, to push knowledge beyond the—still—narrow confines at any given moment, is immanent to human nature. Were this not so, man would still live in caves, being at the mercy of the elements, threatened by wild beasts and by disease. In my opinion there is no reason why this urge to expand knowledge, be it for practical reasons or for its own sake without apparent practical purposes at the moment, should not be applied beyond the earth and into outer space. To obtain results from space explorations which can be practically applied to satisfy human needs that now exist is a welcome byproduct of any scientific project, but does not necessarily, in my opinion, have to be its first motivation."

His Excellency continued:

"Only few nations have the means significantly to contribute to research in the field of space exploration [but] this fact does not justify the monopolizing of the results of such research. I want to make clear, however, that I do not advocate that those nations who are the most active in that respect share the results of their labor with others who sit back and wait until the ripe fruits fall in their lap. Space exploration is a monumental enterprise of all mankind, of such dimensions—and possible consequences—as mankind has never seen before. Therefore it is essential that all nations, which eventually will benefit from the results, work together and assume their share of the burden, according to their abilities and resources. We attach the greatest significance to international cooperation in all areas of space research and exploration and we are prepared to contribute our share, whether through assuming certain costs in financial terms or through the intellectual capacities of our scientists. Outer space is common to all mankind and international efforts have to assure that this remains so.

"In 1980 operation Spacelab will open a new era in space exploration. We are proud to say that experiments designed by Austrians will be among the many research tasks the mission will try to accomplish. We think we thereby demonstrate that even smaller nations like Austria can contribute to meet the tremendous challenge that lies ahead of us in this field.

"I think it is in the best tradition of human endeavor to

meet these challenges head on, for the benefit of all of us and of future generations."

Indonesia

Indonesia, like other Moslem countries, sees the universe as both teacher of humanity and as humanity's destiny; Mr. T.M. Soelaiman, the Educational and Cultural Attaché of the Indonesian Embassy, sees the permanent habitation of space as inevitable; it is both the fulfillment of the Koran and of Indonesian legend. "Indonesian history is full of dreams about flying from Earth. Ramayana is an ancient Hindu story of how people can fly into space."

In 1334 Prime Minister Gajah Mada of the Kingdom of Majapahit swore his "Palapa Oath": "When Gurave, Seran, Tanjung Pura, Haru, Pahang, Dombo, Bali, Sunda, Palembang and Tumasik are united, I shall rest." The oath became a symbol of Indonesian unification, a legend to be fulfilled. In 1976, the promise was kept: President Suharto launched Indonesia's first satellite system and called it "Palapa." The system has integrated the nation as never before, since it creates an electronics communication link among the islands. A second satellite is planned for 1978, making Indonesia the Third World leader in space development.

The Apollo-Soyuz joint U.S.-U.S.S.R. docking gave great hope to the Third World, Mr. Soelaiman noted. But he does not see the world through rose-colored glasses. "Space war is inevitable," he declared. "It is in the human character to be jealous and proud. I expect the future to be like past history." Mr. Soelaiman nonetheless has hope: "Looking at the universe, we are not so proud. We are small in comparison to the universe. We are not all-knowing." So space can restore man's ego

One of 40 earth stations that will carry the massive communications load for the Republic of Indonesia's new domestic satellite system. *Courtesy of Aeronutronic Ford Corporation, Newport Beach, California*

to a proper perspective—we are not gods. If we remember that, Mr. Soelaiman believes we can avoid space war.

Courtesy of APN

U.S.S.R.

His Excellency Ambassador Anatoliy F. Dobrynin of the Soviet Union referred questions of space migration to Y.I. Rakowskiy, Acting Director, Foreign Relations Department, Academy of Sciences of the U.S.S.R. Rakowskiy in turn referred us to N.S. Kardashev, corresponding member of the U.S.S.R.

Cosmonauts Yuri Romanenko and Georgi Grechko at the Salyut-6/Soyuz-27 control panel. The space complex broke the record for space endurance. *Courtesy of Foto-khronika Tass*

Academy of Sciences. What follows is a partial transcript of our exchange:

Q: Your science fiction film *Solaris** is a brilliant movie exploring the possibility of extraterrestrial contact. If the Soviet Union were to make such contact, how might that affect the course of world history?

A: It is apparent that the establishment of contact and even

---

*Solaris is a Soviet science fiction film which deals with a space station on a new planet, Solaris, whose powers are revealed in a series of psychologically suggestive scenes of trauma among the horror-stricken crew.

The dual flags on the panel of the space complex Soyuz-27/Soyuz-28 show the participation of both Soviet and Czechoslovakian cosmonauts. *Courtesy of Novosti Photo*

simply the discovery of the existence of extraterrestrial civilization can radically affect all areas of human activity. It is unlikely we can anticipate "their" active interference with our civilization. However, receiving information about a society which has advanced far ahead in comparison with ours can benefit our evolution and give us a direct indication of the path that is before us. It is highly likely that if contact does take place, the development of humanity would follow the path of rapid assimilation of the knowledge and experience accumulated in another spot in the universe. Of course, how this will take place is impossible to predict, and that is probably why the plot of *Solaris* abruptly comes to an end immediately after the contact with extraterrestrial intelligence.

Q: The Soviet Union has the largest space program in the world. [See Launch and Payload Chart, Chapter 7.] Do your future programs include building permanent communities in space? If so, why should humans live in space?

A: Soviet scientists have a long record of examining the

First tangible accomplishment in international cooperation for humans in space. "Apollo-Soyuz Over the Caspian Sea" by Andrey Sokolov. *Courtesy of the Smithsonian Institution*

possibilities of creating large space settlements. Tsiolkovsky** appears to be correct that in the future a large part of humanity will live in space. This is related to a large number of well-known reasons: the limited natural and energy resources; the growth of population, with the finiteness of habitational surface on the one hand, and with great, practically unlimited opportunities for the development of humanity into space on the other.

Q: In the course of human history, is it now critical to populate space?

A: Apparently, humanity is very close to entry into space; however, this does not mean that efforts to achieve a good life on earth for all people should be stopped.

**Tsiolkovsky, Konstantin (1852-1903), mathematician and aeronautical inventor; 1903, published treatise "Investigating Space with Rocket Devices."

Q: The United States is currently evaluating plans to build human habitats in space. Do you feel the United States should do this alone? In cooperation with the Soviet Union? As part of a future United Nations program? As part of a multinational corporation venture? Or some other arrangement?

A: It seems the best way to go into space is through multinational cooperation. In any case, maximum efforts must be made so that existing contradictions in decision-making regarding the solution of earthly problems will not be transferred into space.

Iran

Mr. Djahanguir Ghobadi of the Iranian Embassy believes that American and Soviet monopoly of space is unhealthy. "The exclusive use of space by the Super Powers increases the possibility of the military use of space. I believe they should meet to establish a framework for the peaceful uses of space and arrangements for the sharing of power with the rest of the world. After they have come to an agreement, they should call for a U.N. Conference on Space." This would be similar to the United Nations Conferences on the environment, food, habitat, and so forth.

Mr. Ghobadi believes this approach is the most efficient, as well as the safest. Until the leaders of space development agree, there is no point in gathering diplomats from around the world to watch the Super Powers jockey for position. Such an agreement would also, he believes, pragmatically reflect the Super Powers' stated beliefs and "show their good intentions for civilian use."

West Germany

Mr. Christian Patermann, Science Counselor of the Federal Republic of Germany, personally views the future of space as "simply a pragmatic question. With the growth in population, depletion of resources and apparent course of evolution, there may one day be human settlements in space. What is science fiction today will be reality tomorrow."

With a background in economics and law, Mr. Patermann sees a variety of institutional arrangements for space migration. He feels joint ventures are to be preferred because of the cost and complexity of space industrialization. Failing that, indi-

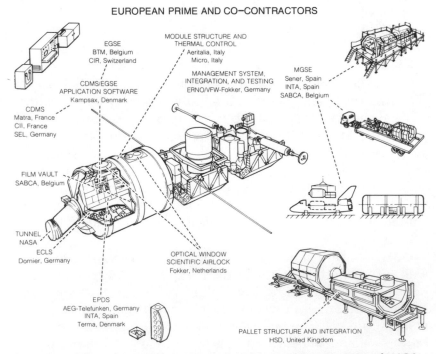

European prime and co-contractors for spacelab and orbiter. *Courtesy of NASA*

vidual nations—like Germany—would expand their space programs and put up their own satellites. The West German firm OTRAG has already invested $30 million in commercial space development, launched its first major rocket on May 17, 1977, from Zaire, and plans to offer launch services sometime between 1979 and 1981. As a diplomat, Mr. Patermann did not want to comment on OTRAG's efforts. Although he finds it unlikely, he does not discount the possibility of Germany's venturing into space alone. Currently West Germany's major attention is devoted to cooperating with the European Space Agency in the development of the Space Lab.

Mr. Patermann was not sure how a ban on war and violence in space could be developed or enforced. His happiest projection was that of the United States, the European Space Agency, and Japan coming to some space migration agreement with the Soviet Union. He also felt multinational space corpo-

Rocket launching site in Zaire for OTRAG, a German aerospace corporation. *Courtesy of Frank K. Wukasch, OTRAG (Orbital Transport and Raketen AG), Stuttgart, Germany*

rations could form something like Intelsat. But nation-by-nation negotiation is the more likely path, Mr. Patermann asserts.

Looking into the far future, Mr. Patermann expects Spacekind and Earthkind to evolve separate ways of life, but not to develop as independent species.

Colombia

Mr. Jaime Lopez-Reyes, Minister Counselor for the Colombian Embassy, looks at space, technology, and progress from a long historical view. "We would like to have some Colombians in space, if for no other reason than to be part of progress. . . .

"We pioneered in air travel. We were the first country of the Americas to have an airline, just after KLM. We didn't know much about airplanes then, just like we don't know much about space technology now. But Germans after World War I came to Barranquilla, Colombia, with small two-seater airplanes adapted as sea planes and began the first airlines in 1919."

Referring to the movie *The Outer Space Connection*, Mr. Lopez-Reyes went on: "We have gold objects that could have been airplane copies and large areas that could have been Pre-Columbian landing ports for space travelers. Where did the Incas learn from? So naturally, we look at things like space travel as progress."

Today, everything is in the hands of the United States and the Soviet Union. "They should invite people to learn or establish an interchange of knowledge to teach people just as people are invited to attend the war college."

He went on to say that he did not feel that the U.N. would successfully lead us into space. Healthy competition between the U.S. and the U.S.S.R. would have the better result: "Competition is always something that has existed in history. Before, it was Spain and Great Britain with Napoleon, or before that, the Greeks and the Romans. It differed in detail, but the competition was there. The man who came up with the idea of the wooden horse in Troy won. Better ideas always come."

Mr. Lopez-Reyes knew John Glenn and Neal Armstrong while he was Counselor General for Colombia in Houston and was proud that a Colombian was an engineer on the NASA Skylab project. Colombia is buying its own satellite from NASA for $102 million; it will be operational in 1980.

Mr. Lopez-Reyes' belief that the U.S. and the U.S.S.R. should compete to help the world get into space just might be a better idea.

Southeast Asia

Or is there a better long-range strategy? One ambassador from Southeast Asia, who asked to remain anonymous, said: "The United States should begin with President Carter's setting his own goals publicly. Next the U.S. should go to its major allies—Japan, Germany, etc.—and develop programs together. As time goes on, new ways will be found for cooperation with the Soviet Union. If and when this becomes concrete, the fourth and final step would be the United Nations involvement. But begin with President Carter, not the U.N."

Meanwhile, the ambassador contended "that the national state system will be so different" at the turn of the year 2000 "that first we must figure it out, then go from there." Today we have 147 sovereign nations; in the future he felt there would be only two sovereign powers—Earth and space, each separate from the other. Hence, Spacekind might well become independent from Earthkind.

Before all those grand plans grow to reality, the ambassador cautioned, the uses of space in relation to earth's needs will have to be as fully understood as possible.

"What's in it for us? Our problems are hunger, disease, poverty, and education. What would happen if there was a moratorium on all space research and development for one year, except in medicine? What will we lose? How far can space assist us in poverty, health, lack of education, and agriculture?

Investing in space is far too much money *unless* applied to
these basic fundamental human problems." The Southeast
Asian ambassador fully understood the value of satellites but
felt that their value had not been fully communicated yet. "In
the best of all possible worlds, the U.N. should do all this, and it
[space] may well catalyze and galvanize the U.N. But it hasn't.
American and Soviet cooperation would be helpful, if it doesn't
become a rivalry. Technicians could specify one specific area
for cooperation and go from there. . . .

    "But first the United States should set its own goals, then
go to the allies to develop goals together, eventually cooperat-
ing with the Soviet Union and finally to the U.N."

Switzerland

    His Excellency Ambassador Raymond Probst of Switzer-
land agrees with the Southeast Asian ambassador's suggested
progression in the development of space. But, he adds, a neutral country will be needed to bring the United States and the
Soviet Union into agreement before a U.N. conference on space
migration could be held with any reasonable hope of success.

    Ambassador Probst holds that extraterrestrial space is ter-
ritory common to all the human species, and that it is beyond
the limits of present national jurisdictions. Ambassador Probst
wrote an article on space law shortly after the launch of Sputnik. He emphasized, "I believe there should be no competition
between states in space, no U.S. colony, no Swiss colony, and
no U.S.S.R. colony. It should be a common enterprise." He
advised against multinational corporations leading the way.
Such competition could lead to conflict.

    So what should be done? "Begin with Carter and Breznev
and a national state with a neutral position to mediate such a

Soviet and U.S. astronauts and space experts holding a conference in Zvesdniy Gorodok (Star City). *Courtesy of Novosti Press Agency*

American Thomas Stafford and Soviet Aleksey Leonov shake hands in Earth orbit just after Apollo-Soyuz docking. *Courtesy of NASA*

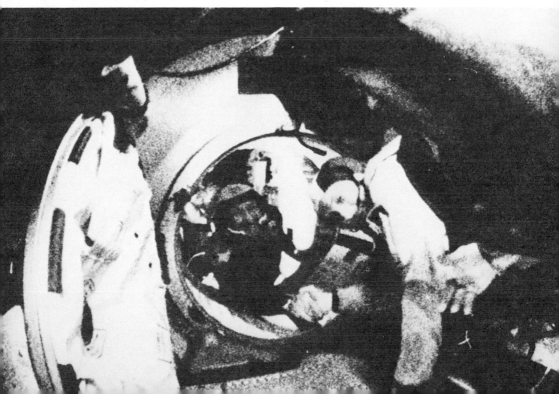

meeting. We will need to create some form of binding arbitration between the Soviets and Americans." The Ambassador indicated that the request to a neutral country must come from President Carter and Secretary Breznev themselves, not from the secretaries of state. A neutral country cannot initiate it, he said, but "We offer the good offices and permanent neutrality of Switzerland for such a meeting."

Looking further into the future, Ambassador Probst speculated about the evolution of humans living independently in space: "When they [Spacekind] feel a bigger loyalty to their comrades than to earth—depending on how long they stay—a new belonging will emerge."

These interviews dispel the idea that the rest of the world doesn't care about the development of outer space. *They do not want to be spectators.* They want to share the costs, responsibilities, and the fruits of space migration. They want in now. And they want to begin the agreements to go.

We could be witnessing the start of the most interesting chapter in diplomatic history. The world is ready to negotiate the future of space migration.

CHAPTER 6

Money Talks: A Future
With and Without Space

Economics of Space

The project manager of a major study for Congress laughed when space was proposed as part of the future economic picture. It was not the first time a misinformed person has sneered, "Why spend money on the moon when there are serious problems of poverty, education, and inflation at home on Earth?"

Let's look at the record. According to the Midwest Research institute, NASA spent $25 billion on research and development between 1959 and 1969. This investment had actually increased the nation's total wealth by $52 billion by 1970. A whopping $181 billion will be returned by 1987.

This investment has stimulated many sectors of the economy, because it causes the contracted companies to increase their own spending. Furthermore, since NASA inventions and designs can be manufactured by anyone, wholly new industries are becoming sources of national wealth and supply.

107

Among their products are digital watches, hand calculators, and new testing machinery.

Net Profit for the U.S. Treasury

What would happen if the American public began to see NASA research and development not as a national ego trip but as a financial investment? According to Chase Econometric Associates, $12 billion added to research and development, $1 billion a year, will add $144 billion to the U.S. economy in twelve years. Every single dollar will return 14.4 dollars. What existing federal program can say that?

After nine years, the wealth of the nation will be increased by $60.3 billion. This will return $11.12 billion in private and corporate taxes to the U.S. Treasury. Thus, from the ninth year the government will receive more money each year than it invests. NASA research will be reducing the national debt! If the resulting tax incomes of state and local governments are also considered, the NASA investment will return a net profit after only six and a half years.

In ten years, this will add 1.1 million new jobs, reducing the rate of unemployment by .41 percent. It's obvious that an increased space effort should add jobs, but it is seldom realized that it also should reduce the rate of inflation by 2 percent. Space R&D contradicts a basic economic theory that increased government spending is always inflationary. Its unique contribution is that it increases the value of goods while reducing their cost. For example, during the fifties and sixties, an office calculator cost between $350 and $500. Because of space research, they are smaller, simpler, more aesthetically pleasing, and much cheaper. A hand calculator can be bought for eight dollars. It has become a highly practical cliché to say, "I'll wait a few years until the cost comes down."

Space research and development has brought down costs of over 3,000 products and services. The first to come to mind are the digital wrist watch and satellite communications. Less obvious is the lightweight fire fighting equipment that saves billions each year in damages. Freeze-dried food can be stored without spending money or expending energy. Manufacturing

Ten-dollar wrist radio developed by the Aerospace Corporation could become the future worldwide CB radio system. *Courtesy of NASA*

costs have been reduced by new alloys, container-free melting, and the process of forming crystals from vapors. Computerized structure analysis saves testing and repair money. Even management planning techniques have been improved.

Space exploration has already paid for itself, and the final total for space's economic contribution will never be completed. It will open up a standard of living impossible to conceive today.

Hot Tip for Investors

The solar power satellite, the first step of space migration, will be a profitable business venture. If work is begun in 1980, a $200 to $400 billion market could be served by the year 2000 or 2004. The competition, nuclear fission light-water reactors and possibly fusion reactors, presently produce electricity at a cost

Thirty-three-foot diameter satellite receiving dish in Orlando, Florida, designed by the American Satellite Corporation, receives facsimiles of *The Wall Street Journal* from Chicopee, Massachusetts, via the Weststar I communications satellite. *Courtesy of Dow Jones & Co.*

of 18.2 mills per kilowatt hour. By 1985, nuclear costs will reach 40 mills. Using O'Neill's approach, the cost of satellite power will be nine mills—half the present cost of nuclear-produced electricity. The U.S. Department of Energy estimates satellite power costs to be 30 mills because of its outdated assumption that Earth materials will be used to build the satellites.

Mark Hopkins of Harvard University has figured the possible return on investment. A benefit/cost ratio of 1.0 means a return of interest but no profit. Any figure over 1.0 shows the number of times costs can increase before the project ceases to be financially worthwhile. Using varying assumptions, the ratio turned out 1.3, 2.1, and 5.2. No matter how you slice the cake, it's a profitable venture.

How much will the first solar power satellite cost? According to Dr. Gerard O'Neill and his colleagues, somewhere between $50 and $60 billion, spread out over the first eleven years of regular use. Hence, the payback would begin the eleventh or twelfth year. This assumes the U.S. Government or a consortium of governments pay the research, development, and demonstration costs over the first several years, expecting 10 percent discounting, and that private investment will take over when risks are lower, expecting a 15 percent profit.

Meetings of private investors have been held across the U.S. and in other countries to figure out how to get in on the future of space manufacturing and migration. Christian Basler, investment counselor, told the American Astronautical Society that he believes a new type of business structure must be developed. His report states that no existing private enterprise structure can raise and manage the capital for full-scale space industrialization. Existing companies cannot undertake the necessary R&D without damaging their present earnings. Antitrust problems would plague them.

Basler proposes the "staging company." As an investment company, it will begin by accumulating capital, investing in companies likely to profit from space industrialization, and spending income on space research and development. The next stage will be attracting firm bids on space systems from the companies that have won previous investments. The capital to

proceed with full-scale space industrialization will be accumulated by public offerings and investment appreciation. When
the national economy has become fully committed to space
industrialization, the staging company will convert to an
operating company.

A Richer Existence

Today there is no generally accepted image of the American or world economic future. As a result, we see many of the
indicators of a civilization's fall—inflation, general anxiety,
falling fertility rates, uncoordinated and contradictory growth,
and general breakdown of traditional values. We just don't
know where we're going.

The currently popular image of the future is the no-growth
society. Its major proponent is the Club of Rome, a powerful
group of European industrialists and scientists who asked
computer scientists at MIT to make a simulation of the world
and, based on current trends in population, consumption, resources, and the like, come up with a long-term projection of
the Earth's future. Their conclusion, reported in *Limits to
Growth*, was that the world is doomed. We were urged to slow
down this global suicide by reorienting civilization toward
small-scale, labor-intensive, recycling productivity.

This concept has its roots in Malthusian theory. Three
hundred years ago, Reverend Thomas Malthus said that a population multiplies without restraint until the environment cannot support its numbers. Growth is naturally checked only by
famine, war, and pestilence, and then the reduced population
immediately begins expanding again. For this law alone,
economics truly deserves the title of the "dismal science."

The return to such dire predictions for humankind is a
predictable overreaction to the problems caused by America's
growth since World War II. This explosion of industry has
polluted our air and water. It has also made Americans
energy-dependent.

But imagine a wealthy Western industrialist flying to the
west coast of Africa and saying, "Stop your growth!" It's like
telling a three-months pregnant woman that she's grown

enough and that's it. Largely because of pressure from Third World leaders, Aurelio Peccei and others of the Club of Rome now say that we must grow, but in new directions. Exactly! We can grow, create less pollution, *and* use fewer resources.

Malthusian theory and the no-growth mandates are wrong, because they are based on the assumption of a closed geography with fixed and finite resources. Today we are using far more technology and proportionately far fewer resources than ever before. Fiber optics carry 100,000 times more communication than conventional copper wire. A chip puts more computer power in one hand than an entire warehouse of vacuum tubes. History shows that we have always created new tools to make those seemingly fixed resources far more effective.

Third World countries will no doubt make great moves toward conventional industrialization; however, the catch-up move will be integration into the Space Age communications system. Through satellites and computers, villagers in Africa

Communications satellite experiment in Kenya, Africa. *Courtesy of NASA*

will receive the best education and medical diagnosis and treatment that Industrial Age money can buy. Farming can be greatly improved through satellite photography.

A future without space technology condemns the Third World to the development gap. A communications-based world will lead to a new kind of wealth distribution. The coming definition of wealth will not be land or material products, but access to information and services. Granted, it is impossible to jump from subsistence agriculture to a communications economy, but it is also unnecessary to follow the same path as the United States—some steps can be skipped.

The no-growth or limits-to-growth, Earth-only future condemns much more than the Third World. It would fail to support the intellectual and cultural growth of civilization. If society becomes convinced that we can live only on Earth with its fixed and finite resources, then we are like animals fighting over one watering hole. At first we will relish every last drop through elaborate and paganistic rituals. Rules will regulate its use. But demand for the water is going to increase.

Scientific breakthroughs in immunity, genetics, and cryogenics may extend life indefinitely. We will clamor for more law and order to protect and justly distribute the water. To enforce these tightened rules, government will become totalitarian. And soonor or later, the response to dictatorship is war—especially in the Third World countries which have never experienced an abundance of wealth.

All things grow or they die. Civilizations—an expression of life—face the same challenge to grow or die. They die when their members have a general agreement that the future will be worse than now. They grow when they agree that the future will be better. The really long-range difference between a future with space and a future without space is simple—without it our prospects are finite; with space our prospects are infinite.

THE HUMAN FUTURE

WITH SPACE MIGRATION	WITHOUT SPACE MIGRATION
Growth	Maintenance
Infinite; no known limits to human existence	Finite; longest future expectancy is 10 billion years
Open	Closed
Increase in human wealth through advanced technology	Malthusian economics with the rise of intermediate technology
Space for earthlings to grow with their curiosity and life extension	Controlled boredom with increasing life expectancy
Peaceful accommodation of growing numbers of political groups and philosophies	Territorial political pressure and conflicts
Pressure for diversity	Pressure for similarity
Continuous sharing of knowledge with continual opening of new frontiers	Continual prioritization of useful research, and defense of local turf
Star Trek	*Brave New World*
Both gravity and zero-gravity living	Only gravity living

CHAPTER 7

Frontiers of
Space Migration

Space Communities—Just Around the Corner

The current NASA plans for humans in space, in our opinion, lack vision. Even if for understandable reasons, NASA's essentially conservative approach has resulted in designs for prototypes that are little more imaginative than, say, those for ecologically advanced garbage cans. They presuppose that life in space will consist almost exclusively of the repetition of mechanical tasks in a mechanical environment.

But the truth is that there are some exciting alternatives to the current NASA plans. In order for them to be conceived of in the first place, a basic premise has to be granted as necessary and possible; the premise itself leaps from an Earthbound imagination just as space migration means a leap from Earth. It is simple enough: *People* migrate to space; it's not just machines out there.

For space migration to be exciting, it has to be made possible for women, children, artists, the whole rich diversity of people that make life worth living. The environment they live

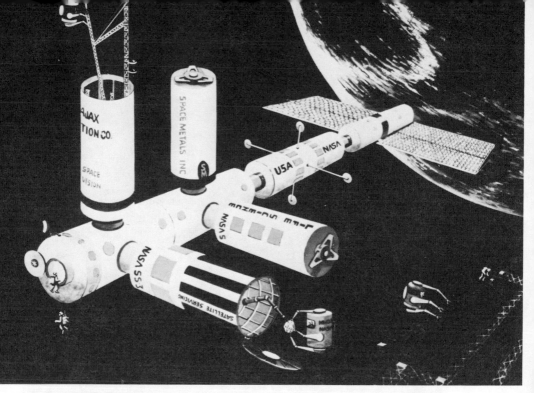

Early concept of a Space Industrial Facility in Earth orbit, having a "garbage can" configuration. This approach to a space station would combine the orbiting laboratory with a construction camp, work station, and industrial complex. Some parts are shown as owned by corporations and others by government. *Courtesy of NASA*

and work in must be at least as interesting as the one they have left. If the environment is spartan and the life repetitive, the early space pioneers may find it no more than interesting for a while, and then rush to get back to Earth. If that happens, the whole space migration effort will be slowed down. But if they enjoy and prefer living out there, then space will become a sought-after place to go, and humanity will find itself rapidly expanding beyond Earth.

Let's take a look at some of the possibilities that make living in space an exciting venture.

Science Fiction Didn't Go Far Enough

The majority of the science fiction of the past has assumed that humans will inhabit other worlds or other planets—that we will build on the moon, on Mars, on Jupiter, and then on other planets of other solar systems in our galaxy.

Construction of solar cells in space with nearby community center. *Courtesy of NASA*

Typical view of lunar station. "Lunar Farside Research Base" by Pierre Mion. *Courtesy of the Smithsonian Institution*

Viking photo of Mars showing pockmarked surface and thin trace of atmosphere above horizon. Mars, the Moon, and other planets are inhospitable, requiring far more expense to make safe than building in free space. *Courtesy of NASA*

But why limit the possibility of life in space only to planets?

Why not make totally new worlds in free space?

And new worlds that can create more new worlds?

Such ideas are not as farfetched as they seem at first.

The first person to seriously propose free-floating space

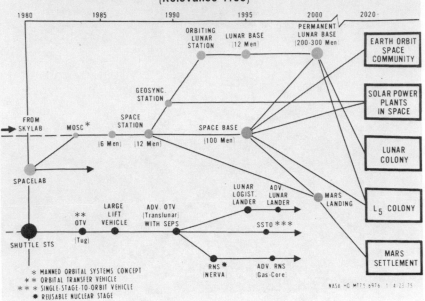

EVOLUTIONARY PATHS TO FAR-FUTURE SPACE ENDEAVORS
(Relevance Tree)

* MANNED ORBITAL SYSTEMS CONCEPT
** ORBITAL TRANSFER VEHICLE
*** SINGLE-STAGE-TO-ORBIT VEHICLE
● REUSABLE NUCLEAR STAGE

NASA-HQ MT75 6976 1 4-23-75

SPACE INDUSTRIALIZATION

1. **PUBLIC SERVICES**
   - INFORMATION TRANSMISSION (Education, medical aid, mail, news services, teleconferences, communications, telemonitoring & teleoperation, time, navigation, search/rescue, etc.)
   - DATA ACQUISITION (Earth resources, crop measurement, food production, water resources, weather & climate, oceans, oil/mineral location, environment, mapping & inventory, etc.)

2. **PRODUCTS**
   - ORGANIC (Biochemicals: Isozymes, urokinase, insulin, etc.)
   - INORGANIC (Semiconductors, large single crystals, high-strength fibers, perfect glasses, new alloys, high-strength magnets, etc.)

3. **ENERGY**
   - SOLAR POWER FROM SPACE
   - FUSION RESEARCH IN SPACE
   - ILLUMINATION FROM SPACE

4. **PEOPLE**
   - MEDICAL & GENETIC RESEARCH
   - SPACE SCIENCES & SPACE-BORN EDUCATION
   - SPACE THERAPEUTICS (Hospital, Sanitarium)
   - SPACE TOURISM (Hotel)
   - ENTERTAINMENT & ARTS

A view of NASA's long-range plan. *Courtesy of NASA*

cities was the Russian Konstantin Eduardovitch Tsiol-kovsky—at the turn of the century. Moving up to the present day, Dr. Gerard O'Neill of Princeton University, along with many others, is actively proposing whole communities in open space. The idea has been aired before the Congress, discussed in professional magazines around the world, and even talked about before huge TV audiences on the Johnny Carson Show.

The concept of communities built in free space has already sparked a number of highly plausible theories and plans.

Professor O'Neill's early suggestion was to build them at what are called L-4 and L-5, points that would remain stationary relative both to the earth and to the moon as it revolves around earth.

Later, in 1976, a NASA study determined that these space cities would be better placed in what is called a 2:1 resonant orbit around earth. This elliptical orbit moves from 200,000 miles from earth at its farthest point to 100,000 miles at its closest. The space community would take about two weeks to complete its orbit.

A 2:1 orbit makes for easier access to the space community from both the earth and the moon than the L-4 or L-5 locations do.

O'Neill's High Frontier

Free-floating cities in space make space a highway, not just the envelope of a planet. It's an exciting idea, it's unique—and it can be started now, using essentially existing and known technology.

The first major product of such a community would be gigantic satellites to collect solar energy. Called solar power satellites (SPS) and weighing about 84,000 tons, these new energy plants would microwave energy to receiving stations on Earth. It is estimated that the cost of electricity on Earth from SPS would be nine mills per kilowatt hour, or half the cost of current nuclear-produced electricity. This assumes that the materials to build them come from the moon.

O'Neill suggests a simple mining operation on the moon that involves no processing. Instead, lunar matter would sim-

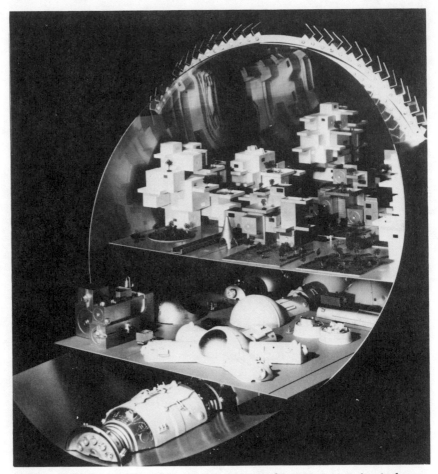

Artist's model of the interior of a torus, or outer ring of a space community. At the very top of the cylinder are mirrors that reflect sunlight into the community by baffles that prevent cosmic rays from entering. *Courtesy of NASA*

ply be "thrown" to a particular point in free space, there to be processed by a plant near or at the space community. The thrower, or "mass driver," is an electrically powered magnetic accelerator with laser guidance. Raw lunar material is put in a bucket. The bucket is magnetically levitated inside a series of coils. As the magnetic field is altered, the bucket is accelerated. It is a linear synchronous motor.

A small prototype of the mass driver has already been built

by students at MIT. It accelerated one pound of matter at 33 gravities to 81 miles per hour in eight feet in 0.11 seconds. O'Neill himself is working on one for 1,000 gravities to reach 350 miles per hour in just four feet (1.2 meters). Larger mass drivers could be used to get material from low orbit to high orbit, from high orbit to geosynchronous orbit, from the moon to L-2 or other points in free space.

One could imagine in the more distant future a gigantic mass driver for space communities. It could accelerate very slowly at first and then gradually reach extraordinary speeds to propel entire space communities out of the solar system. Kind of a "man shot from cannon" act, but on a scale undreamed of by Barnum and Bailey at their wildest.

However, Jesco von Putkammer of NASA's Advanced Programs claims that the mass driver may well be useful for lifting lunar material, but feels the reaction mass "engine" or mass driver would be impractical in free space. He disagrees with its use either in low orbit or for asteroids.

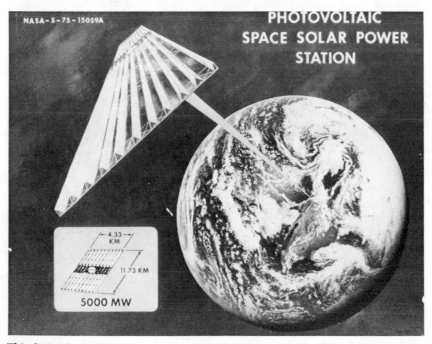

This design by O'Neill and Associates for a 5,000-megawatt station is 4.33 kilometers wide and 11.73 kilometers long. *Courtesy of NASA*

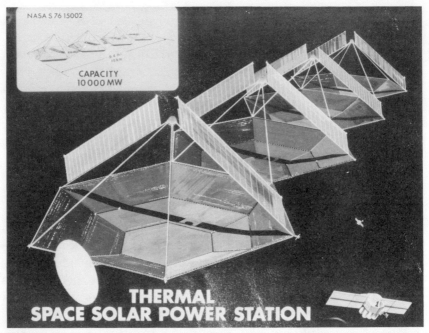

THERMAL SPACE SOLAR POWER STATION

This Boeing 10,000 megawatt station is over 15 kilometers (9.4 miles) long. *Courtesy of Boeing, Inc.*

Artist's conception of a lunar mining facility. *Courtesy of NASA*

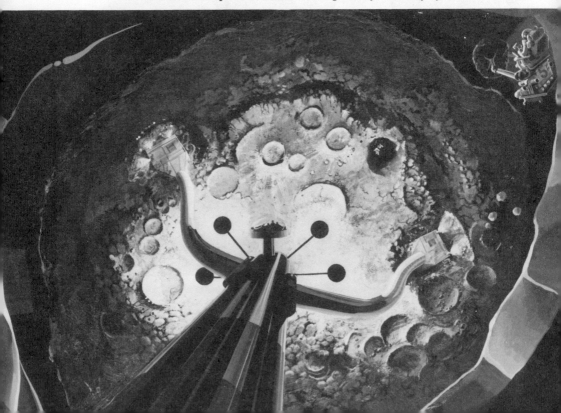

Lunar materials are processed and manufactured in this artist's conception of an orbital facility. *Courtesy of NASA*

The first space community, as envisioned by O'Neill, would spend 60 percent of its time producing solar power stations to orbit Earth and microwave their collected energy to Earth. The remaining 40 percent of the time would be spent making another partly independent community.

The last conception is what makes the idea of space communities unique and exciting: the idea of creating a community that will create other communities that are free offsprings from Mother Earth—to be different from Earthkind, and even different from each other.

The basic plan, as explained by O'Neill in the February 1976 issue of *The Futurist*, proposes:

Lunar mass driver, one section in foreground and other sections stretching back off the horizon. *Courtesy of NASA*

1. To establish a highly industrialized, self-maintaining human community in free space, at a location along the orbit of the moon called L-5 [2:1 resonant orbit will more likely be used], where free solar energy is available fulltime.

2. To construct that community on a short time scale, without depending on rocket engines any more advanced than those of the space shuttle.

3. To reduce the costs greatly by obtaining nearly all of the construction materials from the surface of the moon. (Far cheaper than taking materials from the Earth. If asteroids were used instead of lunar material, costs drop another 50 percent.)

4. At the space community, to process lunar surface raw materials into metals, ceramics, glass, and oxygen for the construction of additional communities and of

products such as satellite solar power stations. The power stations would be relocated in synchronous orbit about the Earth, to supply the Earth with electrical energy by low-density microwave beams. (A 2-4 GHz beam attenuates from 2 to 10 percent depending on rainfall. Transmission efficiency is over 90 percent and total system efficiency is 54 percent.)

5. Throughout the program, to rely only on those technologies which are available at the same time, while recognizing and supporting the developments of more advanced technologies if their benefits are clear.

From the decision to build solar power satellites until the first one is powering light bulbs on Earth as a regular service will take about twelve years. A solar power satellite could be

First mass driver, designed by H. Kolm and G. K. O'Neill, and built by group of students at MIT. It was started during the January Independent Activity Period, 1977, and finished in time to be demonstrated at the Princeton Symposium on Space Manufacturing, May 1977. It accelerates a one-pound bucket at 35g, reaching a speed of 90 mph in six feet. Left to right: Dr. William Wheaton (post-doctoral fellow in X-ray astronomy), Kevin Fine, Bill Snow (both graduate students in the Aero-Astro Department at MIT), Jonah Garbus, and Eric Drexler (both undergraduate students in Aero-Astro); back to camera: Graham Chedd, producer of the two NOVA programs *One Small Step* and *The Final Frontier. Courtesy of Henry H. Kolm*

Mass driver being loaded onto C-130 Hercules for trip to California, where it was demonstrated on the occasion of the first flight of the space shuttle *Enterprise* in August 1977, arranged by former astronaut Rusty Schweickart. *Courtesy of Henry H. Kolm*

built within six years. By comparison, it takes ten years from decision to completion of a nuclear power plant. The following chart, prepared with the help of Mark Hopkins and Gerald Driggers, is Dr. O'Neill's preferred schedule:

1978–1980	Study environmental, economic, social, and national security impacts. Ground-test hardware.
1980	Decision to go or not to go.
1981–1984	Apollo-scale program to develop hardware and test in orbit.
1985	First landing of mining equipment on the moon and beginning construction of small lunar base.
1987	Lunar mining operation begins sending materials to a point in free space for processing.

1988-1991   First chemical processing plant in free space producing over half-million tons of material per year. Test solar power satellite parts. Workers living in apartments made in used liquid fuel tanks from the space shuttle.

1992   First solar power satellite (84,000 tons) comes on line for regular use at a value of $20 billion, producing 10 gigawatts, or 10,000 megawatts, in a geosynchronous orbit 23,000 miles above Earth.

1993-2004   First Earthlike habitat for 100,000 people. Rapid production of solar power satellites, yielding a total of from 200-400 gigawatts per year, at a value of $200-$400 billion.

Costs could be cut by using asteroids that are near the Earth. There are possibly as many as 2,400 with a diameter over one kilometer that will collide with either the earth or moon over the next 100 million years. These asteroids could be sent to the manufacturing station in free space by a mass driver, which is two to three times cheaper than a chemical rocket tug. If asteroids were to be used in this way, it could eliminate the need for a lunar mining base entirely and cut the project cost in half.

Brian O'Leary, assistant to Dr. O'Neill and former astronaut, goes even farther. He has proposed that such asteroids could be broken up, covered with a special foam, sent back to Earth's oceans, recovered, and their iron and nickel content extracted for use here on Earth.

Unfortunately, the Department of Energy and NASA have not taken the use of extraterrestrial materials—lunar or asteroidal—into account in their studies of solar power satellites. Instead, they assume all materials will come from the Earth. This makes the project ridiculously expensive; the cost is estimated at over twenty times more than O'Neill is suggesting.

The first community in the 2:1 resonant orbit around Earth may look like the wheel-shaped city similar to the one in the film *2001: A Space Odyssey*. If begun in 1980, O'Neill

The 10,000-inhabitant torus space community. *Courtesy of NASA*

claims that it could be complete by 1992, able to produce things for the benefit of Earth.

The community would employ ten thousand people, working and living in an Earthlike environment complete with trees, birds, and rivers. The inhabitants would grow their own food. Each crop would have appropriately controlled temperature, humidity, weather, gravity, and atmospheric pressure; even the number of "daylight" hours and seasons of the year would be controlled. It is estimated that 111 acres will feed those ten thousand people.

Bringing nature into space in this way may well prove more difficult than presently anticipated. Ecological balances are not easy to ascertain and maintain, and we may see some rather bizarre overgrowths or variations of certain species until the situation can be stabilized and corrected.

But the possibility of error is reduced to almost no consequence compared to how much we stand to learn about nature by trying to duplicate it in space. One of the most hopeful pictures of the Earth's future is one in which both high technology and the natural environment are in close to perfect balance. Space communities can be brilliant experimenting grounds to learn how to accomplish that—and bring it back down to Earth.

The Soviet Union's recent six-month ground simulation of an orbiting capsule tends to prove beyond doubt that the possibility of such self-sustenance is not fantasy. The capsule held three people. They derived their oxygen through recovery from atmospheric carbon dioxide, their water from a photosynthetic regenerative process; and they produced some grain and vegetables as food.

I.I. Gitelson from the Soviet Academy of Sciences claimed that this "established beyond question that a small, steadily operating, essentially closed system of 'substance-turnover' involving man is quite feasible." These techniques were further refined during the Soviet record-breaking ninety-six days in orbit.

Concept for a city of 10,000 that emerged from a 1976 study of space manufacturing at NASA's Ames Research Center. A nonrotating spherical shell made of slag shields the workers from rays. *Courtesy of NASA*

Dr. O'Neill's second generation space cities may look more like giant cylinders than those of the first generation. Early ones would be as small as 3,280 feet long and 328 feet wide and employ ten thousand people, while later models could be nineteen miles long and four miles in diameter and employ several million people.

Large-scale production is relatively easy in space. There is no gravity, vacuums are simple and inexpensive to produce, energy is cheap and constant, and the work environment can be controlled and made more predictable. Many of these techniques were apparently successfully tested during the 96-day Soviet orbit.

On Earth, just as one example, tremendous effort goes into lifting up, settling down, or pushing large objects. In space, it's a breeze.

Ideally, in the future all major industrial production will

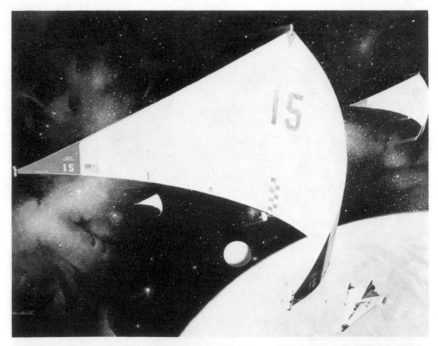

Original concept of solar sail by Thomas Turner, corporate Vice President of Fairchild. Painting by Robert McCall, "Space Sail of the Future." *Courtesy of the Smithsonian Institution*

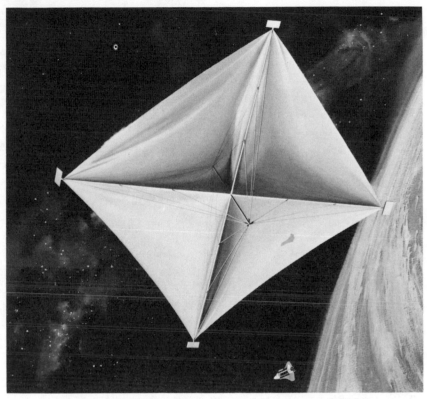

The design for solar sails is still evolving. *Courtesy of NASA*

occur in space; that would allow the Earth to breathe free, clean itself up, and become a veritable "Garden of Eden."

There are a variety of ways to live in outer space: asteroids hollowed out for living inside; interconnecting domes on the moon and other planets; a Mars atmosphere changed so it can contain Earth life. There may be artificial planets made up of gigantic spheres with propulsion systems for intersolar-system travel, or gigantic sails to be pushed along by solar energy.

Other Space Cities

Dr. Peter Lizon, Director of the Graduate School of Architecture at Catholic University of America, believes that Dr. O'Neill's designs are too vulnerable and Earthlike. While teach-

Tetrahedron space city with internal sections that rotate independently. *Courtesy of Dr. Peter Lizon's Design Studio, University of Tennessee*

ing at the University of Tennessee, he and a team of students developed alternatives to O'Neill's centralized cylinders and torus space cities. Rather than the entire city rotating, Lizon's group decided it was safer to compartmentalize or modularize a stationary city, with subsections rotating to produce the required artificial gravity. The tetrahedron was chosen for the overall shape of the city since it is the most stable solid. Keith Critchlow, in his *Order in Space*, put it simply: "The tetrahedron, minimally structured, is the strongest of the fundamental solids, being most able to resist external forces from all directions." A tetrahedron balances four planes equally, while a cylinder is inherently less stable since it centers on one axis.

Lizon feels each subsection of his tetrahedron city should

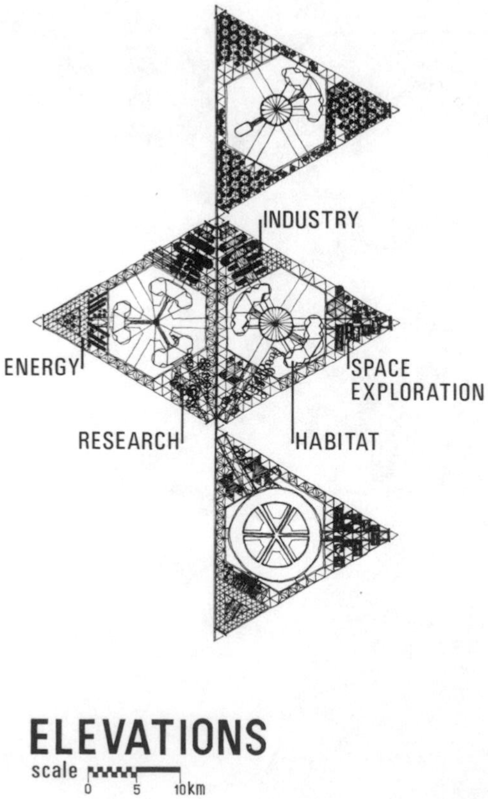

ELEVATIONS

scale 0 5 10km

Tetrahedron cities could "grow like crystals" to enormous size. *Courtesy of Dr. Peter Lizon's Design Studio, University of Tennessee*

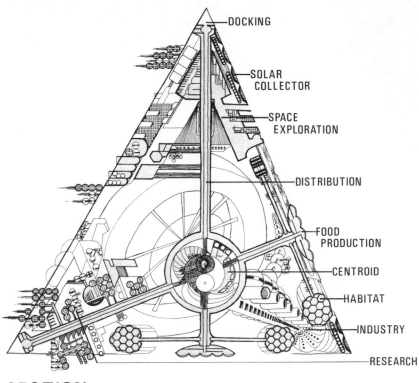

DOCKING

SOLAR COLLECTOR

SPACE EXPLORATION

DISTRIBUTION

FOOD PRODUCTION

CENTROID

HABITAT

INDUSTRY

RESEARCH

SECTION

Cross-section of tetrahedron city showing functional areas. The habitat is located in four rotating gyros placed in the four sides of the form. *Courtesy of Architectural Design Studio of Dr. Peter Lizon, University of Tennessee, 1975*

be a small city unto itself. The tetrahedron is envisioned as twenty-two kilometers long on each side. The smallest grouping might be 105 people, fitting to larger units of 25,000 and 100,000. The social order and the environmental design should be contiguous. The relative independence of each subsection avoids the possibility of one malfunction affecting the entire system; the lights went out all over the northeastern section of the United States because of a malfunction of one little section within the entire electric grid.

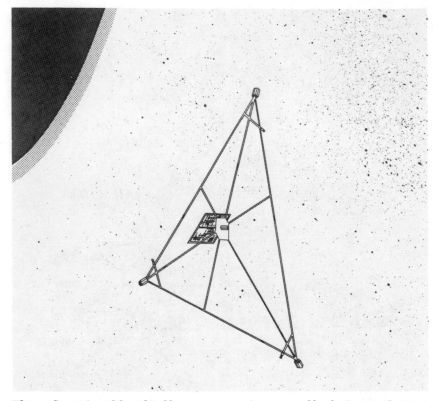

The configuration of the orbital human community proposed by the Rosens. *Courtesy of Stan and Lisa Rosen*

Dr. Lizon's designs are in no way worked out in the detail of O'Neill's; however, they offer an alternative to the central rotating city. The tetrahedron city can add sections as they are needed.

Stan and Lisa Neufeld Rosen propose using the leftover liquid fuel tanks of the space shuttle. A relatively simple orbital human community could be constructed in 1990-2000. Orbiting the earth at an altitude of about four hundred miles and at an inclination of 28.5 degrees to the equator, the community would be resupplied by the space shuttle at regular intervals. Possible activities include materials processing in a range of gravity fields, from zero to one-g; construction of large structures; servicing of Earth satellites and checking out

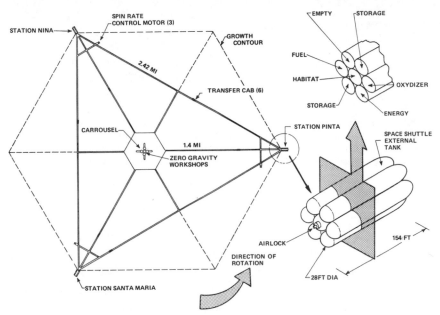

Plan view of the Rosens' base. *Courtesy of Stan and Lisa Rosen*

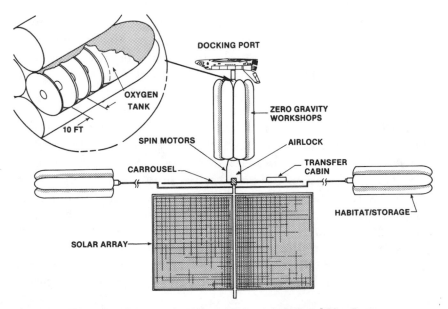

Side view of the Rosen plan. *Courtesy of Stan and Lisa Rosen*

One day in the distant future, space communities might look like the double helix DNA. We have modeled highways after the human circulatory system; maybe DNA, the design of life, is a good model for an entire space civilization, as suggested by John Marsh and Isaac Asimov. *Courtesy of National Institutes of Health*

planetary probes; observation of Earth weather and natural resources; astronomy; and use as a resort for space tourists.

Dwelling and working stations would be built from clustered space shuttle external propellant tanks. Dwellings would be connected by trusswork and would revolve around a central zero-g workshop and docking port to produce the artificial gravity necessary for long-term living in space. Continuously facing the sun, solar cell arrays near the workshops would provide power for the base.

The Rosens would like to call the original three stations the *Niña, Pinta,* and *Santa Maria,* to commemorate the five hundredth anniversary of Columbus' voyage. The dashed line shows how additional tank clusters could be added to meet growth requirements. Spin rate control rocket motors near the periphery provide initial spin and prevent slowing of the rotation rate as mass is transferred outward from the center. Elevatorlike transfer cabins allow movement between dwelling and working stations. The enlargement shows a typical station cluster; the cutaway indicates possible uses for each tank.

The side view of the base shows how the docking port at the top of the nonrotating workshops allows transfer of materials from the space shuttle. Motors at the bottom of the workshop rotate a carousel in synchronization with the base. For transfer to or from the zero-g workshop, the carousel slows its spin rate to zero. The cutaway shows how typical workshop and habitat tanks may be divided into working and living areas.

The vessel originally used for liquid oxygen would provide storage volume.

Imagination Knows No Bounds . . .

A duplication of DNA structure has been suggested as a space community habitat. John Marsh, fourteen years old at the time, proposed a double helix at a recent conference on space habitats, with a radius of 1.59 miles. Isaac Asimov suggests that new constructions be added at one end of the space habitat while older, perhaps outdated structures are removed from the other end, making the entire community obviously move toward continual improvement. Such a community would make a unique calling card for any passing extraterrestrials!

CHAPTER 8

What Will We Do When We Get Out There?

Aside from generating major industrial production for Earth and for new cities in space, Spacekind will be busily acting out through trial and error the next step in evolution—at least culturally, and possibly past that. In fact, one space design conference wanted to name the first space community "Chrysalis," the stage of metamorphosis between the larvae and the butterfly.

The world in general will have learned to cooperate by then, perhaps even to unify, and it is likely that the inhabitants of space cities will represent an international cross-section of humanity. The signs of such cooperation are already emerging. Apollo-Soyuz, the joint U.S.-U.S.S.R. Earth-orbital space station, and the recent Soviet spaceflights with Czechoslovakian and Polish crew members, constitute clear signs of what is coming. As this book is being written, the United States and the Soviet Union are discussing cooperative ventures involving the Salyut and the space shuttle. And China has made inquiries about the possibility of their use of the space shuttle.

It is inconceivable that such worldwide cooperation will not have grown by 1990, in time for truly international space communities.

Imagine being a teacher of a class of twenty young children from Africa, North America, China, Brazil, France, and the Soviet Union in the year 2001! What will be taught at such a school? Will there be classrooms at all?

The old question "Education for what?" takes on staggering implications. Will children need to be taught to communicate with plants and other animals in preparation for nonhuman intelligent contact? After all, the search for other intelligent life in the universe is one of the primary drives of Spacekind.

What kind of recreation will be available? Microwaved coverage of the Army-Navy football game? Electronic sexual affairs with Earthkind? Sexual playpens with spherically padded walls and a large water-filled ball as a bed?

Over the next twenty years biofeedback technology is expected to become far more sophisticated and micromin-

The living accommodations and office of a crew member. They consist of a cylindrical liquid hydrogen tank of the shuttle. These modules would be clustered to form the first human habitats in space during the buildup period of space manufacturing. *Courtesy of NASA*

iaturized. It's far from fantasy to imagine eyeglasses, hats, pants, and other articles of clothing equipped with electronic stimulators and sensors; these would transmit impulses, messages, even sensations via small microwave units to one's lover in a kind of electronic sexual affair. If one were feeling a little lonely or depressed, coded signals could be sent out until another person, wearing the same units and feeling sympathetic, could respond for a secret moment of blissful contact. When being near the one you love is not possible, such electronic means of contact may well become extremely attractive substitutes. Those of Earth might even get some "sensation" of what it's like to experience sex at zero-G—one very direct way of closing a communications gap!

Bruce Murray of the Jet Propulsion Laboratory suggests solar sailing as space recreation. Designed by Tom Turner of Fairchild Industries and improved by Eric Drexler, a student at MIT, the solar sailboat would be propelled by photons striking a large sail of three hundred or more feet. What a Space Olympics event that would be!

Dr. Isaac Asimov suggests that lunar colonists would be great trampoline artists. The moon's gravity, one-sixth that of Earth's, would allow them to fly high and wide for a long time.

Holographic laser art between floating communities at L-5 and the moon could be breathtaking.

The possibilities for communication, art, entertainment, and recreation are almost literally endless.

Andrew Weil, in his book *The Natural Mind*, claims that all cultures have ritualized methods for altering consciousness, from peyote to meditation to cocktail parties. How will the inhabitants of an intercultural space community alter their consciousnesses—that is, beyond the already mind-altering experience of three-dimensional living in gravity-free environments with the Earth, moon, and solar eclipses as normal vistas?

Of course, the purposes of the early space communities will be primarily mining the moon, manufacturing solar power stations, carrying out basic and applied medical and scientific research, performing exploratory astronomy, improving com-

munications, mapping Earth's and other planets' composi-
tions, producing food, manufacturing drugs—and tourism.

Tourism?

Certainly.

Whenever a social experiment takes place, such as Paolo
Solari's Archosanti or Antioch College's graduate schools,
tourists flock to visit. And space communities are without
question more fascinating than the usual run of tourist attractions.

William Brown and Herman Kahn produced a study for
NASA, *Long-Term Prospects for Developments in Space (A
Scenario Approach)* (NASA CR-156837). Their projection for
space tourists:

Year	2000	2025	2050	2075
Pessimistic	0	0	20	1000
Moderate	20	600	10,000	200,000
Optimistic	600	100,000	10^7	10^8

Pat Gunkel, former member of the Hudson Institute, gives a
different estimate in his *Future of Space:*

Year	2000	2025	2050	2075
Probable	100	100,000	10 million	100 million
Optimistic	100,000	100 million	400 million	1 billion
Possible	4 million	Everyone & newborn	10 trillion	New kind of life

Whatever estimate is accepted, whatever new ones are
made, one fact seems very obvious:

We'll be out there.

Life and Death and Recycling

Time spent in low or zero gravity is expected to work wonders for those with heart problems. It may turn out that gravity-free living can slow down the aging process. The differences between Earthkind and Spacekind may shed light and add unimaginable dimension to life-extension medicine. And death? Should people be sent back to Earth, cremated, used as parts for the living, or as fertilizer or food as in the movie *Soylent Green*?

No doubt uniquely appropriate answers will be found as new views of reality and religion develop.

The Need for Social Planning: Various Facets

Sociologists and anthropologists are now beginning to recognize the pressing need to involve themselves seriously in determining the best type of social structure and cultural framework for the inhabitants of space communities.

These communities will reflect a uniquely high degree of integration and interdependence between technology (basic life-support requirements) and the realities of economic dependence while cultivating political independence.

At a recent conference of the American Anthropological Association, Arthur Harkins and Gary Hudson of the University of Minnesota emphasized the importance of technology as an instrument for man to manipulate and redefine the nature of "(1) the present, and (2) the future possibilities open to him." They referred to this concept as "technologizing" the best cultural or social institutions and a workable process for providing a "values-focused assessment of human space colonization alternatives."

Brian O'Leary, former astronaut and physicist associated with Gerard O'Neill at Princeton University, stressed the likelihood of the ultimate socioeconomic independence of space communities from Earth. He emphasized that the political, cultural, and psychological implications of a space community competitively producing solar energy "are enormous, and require the early development of a framework for social planning—for example, the degree of autonomy of communities which may eventually provide the bulk of energy supply to Earth."

Robert J. Miller of the University of Wisconsin put his finger squarely on the general nature of the problems to be resolved in determining the basic cultural characteristics of a space community. First to be considered are the physiological responses to a synthetic and alien life-support system. Second, and just as important, is the issue of control by or dependence on Earth. The issue of control is, in turn, "directly related to the technologies involved and the [selected characteristics of the] population using them." For example, "waste disposal, food supply, habitations (design of which . . . affect matters of privacy, social distancing, etc.) . . . must be provided."

Luther P. Gerlach of the University of Minnesota emphasized the likelihood that life in space communities necessarily means greater control over the inhabitants' interactions with their biophysical environment. He then went on to pose the following critical questions for resolution:

What are the sociocultural costs of such control? What will their effect be on the technoeconomic and sociopolitical integration? Will such control over biophysical evolution require even greater control over the lives of individuals? "Some will see [this] as a leap into escapist fantasy," he concluded, "and others as promising even greater technocratic and ecological risk."

Dr. Magoroh Maruyama warns of the possibility of what some science fiction writers have called "space senility." In a totally controlled environment, one could eventually come to believe that all reality exists within one's own imagination and mind; that nothing exists beyond what one immediately perceives. Maruyama suggests this is preventable by: (1) placing large objects outside the immediate environment, some distance away, simply to be looked at—they literally help supply perspective; (2) allowing unpredictable situations to grow to visible proportions; and (3) experiencing growth in plants, other animals, and children.

In short: A rich diversity of people, plants, other animals, and of the physical environment in general maintains the psychologically necessary sense of the movement, changeability, and unpredictability of life. It creates the imaginative tension between the known and the nonthreateningly unexpected that helps keep personality on an even keel.

Roger Wescott of Drew University took strong issue with NASA's prevailing policy of selecting "space athletes" for manned programs. He feels that maximum "diversity among space stationeers" is to be much preferred, and referred to the possible "admissions policy" for space communities as an "exciting, if demanding, process of ethnogenesis, or culture formation."

Extraterrestrial Contact

Sir Arthur Eddington, a British astronomer long interested in the internal structures of stars and the fundamental constants of nature, tells us that "10^{11} (100,000,000,000 or 100 billion) stars make a galaxy, 10^{11} galaxies make a universe." That's 10^{22} or 10,000,000,000,000,000,000,000 star systems in the known universe. According to the Harvard biologist George Wald, between 1 to 5 percent of the stars in our galaxy could provide the conditions for planetary life, another way of saying there are 1 to 5 billion chances for life in our galaxy. That's at least 100,000,000,000,000,000,000 chances for life in the universe.

The facts—and the sheer odds—all add up to its being simply more reasonable to assume that there is intelligent life in the universe than it is to assume we are the only intelligent life among 10^{22} stars. As a result, on some future day, we may find ourselves confronted with nonhuman, nonsolar-system, and possibly even non-Milky Way intelligent life.

What will our religions have to say about that? The question is not so improbable as it may first seem. Religious institutions play a large part in our lives; religious leaders are very influential in society even if people don't go to church very much.

As proof of that, in 1972 NASA sponsored a symposium at Boston University. Its purpose was to discuss the status of theology in the space age. What follows is a summary of the major speakers' comments on the issues of extraterrestrial

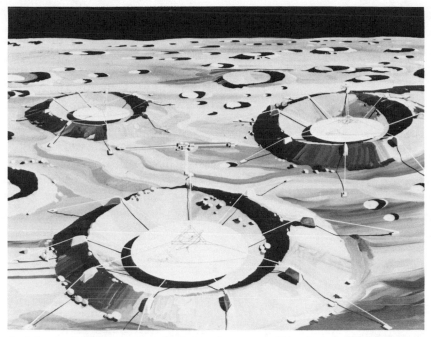

Artist's conception of receivers built into the moon to listen for extraterrestrial contact. *Courtesy of NASA*

communications and their religious consequences; we have added a brief comment. The importance of such a panel being convened should not be underestimated.

DR. GEORGE WALD Harvard University Biologist	Contact with advanced extraterrestrial intelligence would destroy human dignity. A free ride on the advanced extraterrestrial knowledge starts the decay of human culture.
COMMENT:	*Dr. Wald forgets that interaction between technologically different cultures here on Earth provided the basis for culture hybrids and scientific progress. Fear of the unknown insults the inquiring human spirit.*

DR. ASHLEY MONTAGU Rutgers University, Anthropologist and Social Biologist	Contact with advanced extraterrestrial intelligence is unlikely since humans would be perceived "as a highly infectious disease that is best quarantined from the rest of the universe."
COMMENT:	*Dr. Montagu insults Christianity and other anthropocentric theologies as fraudulent opiates for large segments of the public. This view could create a negative self-fulfilling prophecy.*

DR. KRISTER STENDAHL Dean, Harvard School of Theology	Contact with advanced extraterrestrial intelligence and support for science and technology helps God's world get a little bigger and we get a somewhat more true view of our place—and our smallness—in His universe. Scientific methodology is a critical instrument to good theology.
COMMENT:	*Dr. Stendahl demonstrates a wisdom that was lacking in theology during the days of Bruno and Galileo; but he goes farther to assume theocentrically that "If there are any people who have intellectually and emotionally trained themselves for dealing with life beyond Earth, I [Stendahl] guess they would be the theologians, with their angels and archangels and all the company of Heaven."*

DR. PHILIP MORRISON
Massachusetts Institute
of Technology Physicist,
Educator, Philosopher

It is important to attempt communication with extraterrestrial intelligence because it will advance the human condition. Gradually, over lifetimes, the human intellect or spirit will mature and could learn how the extraterrestrials could "fashion a world in which they could live, persevere, and maintain something of worth and beauty. . . ."

COMMENT:

Dr. Morrison emphasizes the importance of our actually doing the search for other cultures and values. He said even if "we do not find that our counterparts exist somewhere else . . . [it] would give us even a heavier responsibility to represent intelligence in this extraordinarily large and diverse universe."

DR. CARL SAGAN
Cornell University
Astronomer and
Exobiologist

Contact with extraterrestrial intelligence is a political-economic problem. Several decades of radio telescope "listening for signals from space cost the equivalent of one day's funding of the Vietnam War." Even "the cost forecasting errors on a single weapons system come to more than a decade's exploration of the entire solar system."

COMMENT:

Dr. Sagan's moral indignation that we spend more to destroy old worlds than to seek new worlds peacefully in space is to be supported.

The possibility of extraterrestrial contacts raises other questions not easily answered.

Harvard biologist Dr. George Wald asks: "To what use should mankind put knowledge and understanding gained from contact with advanced extraterrestrial civilization?" Our early use of the dolphins trained to communicate and learn from humans was to send them to Vietnam for military purposes. Christianity, as a representative human-oriented theology, asserts that all levels of intelligence are uniquely integrated toward what self-styled space philosopher Earl Hubbard refers to as the "creative intent." Surely Vietnam dolphin duty did not serve the creative intent. Dr. William Hamilton, author and dean at Portland State University, asks, "Are there any atheists in intergalactic foxholes?" This is not unlike the question implied in *Close Encounters of a Third Kind*: Can there be spiritual contact with extraterrestrials?

If we are to populate space we will have to recognize and accept the possibility of extraterrestrial contact. Such contact will have consequences beyond our current imagination on all levels of human experience.

Are we ready for it? Are we "mature" enough to accept it? Is our theology prepared for nonhuman intelligence? Can the dangerous games of international power be dropped in the face of extraterrestrial power?

Discussions on the topic happen more and more frequently around the world, even if they reach no firm conclusions. Some theologians talk about it, as we have seen—if a little nervously. And informal proposals have been made to Soviet officials to develop procedures for U.S./U.S.S.R. contact with extraterrestrials. That's all a good start. Maybe, just maybe, we might be ready to meet alien life when contact happens.

Is There Room for Individualism in Space?

Civilizations and cultures have been traditionally viewed as maintaining themselves through time by supplying a sense of both stability and permanence by the building of "things"—monuments or beliefs—that are understood by the builders as lasting beyond their lifetimes. This is reinforced by generally accepted rules of social behavior.

Individualism might be defined as the belief that the final authority for a decision affecting the individual lies in one's self; and that authority is to be honored in and by others if it harms no one else.

Given these definitions, is individualism compatible with civilization in space? The answer is "Yes!"

Certainly, clear rules of social behavior will be established. But instead of material monuments and ideologies, what will characterize space civilizations will be the continual flow and evolution of thought and knowledge, shared by all, connecting progeny with progenitor.

Anticipatory Democracy (edited by Clement Bezold) documents the basic truth that the degree people are involved in a decision is the degree to which the decision is accepted; and the degree to which decision-makers are involved in a participatory process is the degree to which conclusions will be implemented with ease and speed. Such intelligent management should be employed in the development of space cities as well as their maintenance. This approach strengthens the role of individualism in a highly cooperative system.

From this perspective, individualism is essential to a space civilization. But what happens as time passes, and individual growth and demands become different, and disagreements occur over the purpose of the community? In anticipation of this, alternative and independent communities could be built to suit those new and evolving purposes.

Individualism need not be suppressed in favor of community or cultural survival. It needs, in fact, to be nourished for the sake of the continuation and vitality of the space communities.

Refreshing the Human Spirit

The container Earth was never seen as a limit until we populated the globe and set up global communications systems. But now we have, and the pressure to "get out" is mounting—slowly at first, perhaps, but with increasing momentum to push us on to new frontiers.

Captain Stan Rosen found that an altered state of consciousness was a normal and typical reaction of the astronauts
to space flight. In his "Mind in Space," published in the U.S.
Air Force *Medical Service Digest* of January-February 1976, he
wrote that astronauts were reluctant to mention such experiences at first, fearing the flight surgeon might ground them as
psychologically unfit or NASA administrators might ridicule
them. But some started speaking up. Ed Mitchell said, "Something happened to me during the flight that I didn't even
recognize at the time. I would say it was an altered state of
consciousness, a peak experience, if you will." Alfred Worden
wrote in his book, *Hello Earth: Greetings from Endeavor*, that
space flight "changed [my] entire view of reality." James Irwin
said, "I felt the power of God as I'd never felt it before"; he now
works for the High Flight Foundation promoting spiritual
space consciousness.

The immensity, beauty, and grandeur of spaceflight was
dramatized by the early astronauts' exuberance. It opened new
doors of perception, raised new questions never dreamed of
before. Spaceflight will give space travelers a chance to see
themselves from a different and elevated perspective.

The psychologist Abraham Maslow developed a hierarchy
of needs common to all people: physical safety, esteem, love
and belongingness, and self-actualization. Possibly a future
psychologist will discover that a feeling of harmony with the
universe is the next need after self-actualization. If so, space
migration will be considered even more as a necessary experience for a healthy human spirit.

When living in space becomes a true alternative to earthly
living, those who choose challenge over security, the unknown
over the known, will join the space migration not simply as a
challenging alternative, but as a pivotal requirement for survival.

Growth

The resistance of civilizations to change is obvious and
natural, just as it is predictable right down to the simplest
living biological specimen. But resistance to change is not

Large-scale space construction of experimental solar satellite power station. *Courtesy of NASA*

necessarily resistance to growth. The formula frequently serves as a braking mechanism to force environmental changes in an organism—be it a viral strain or a civilization—into *integrated* growth. If there is no change, the growth is replication and not evolution.

If an organism's growth pattern is simply replicative, it can be considered cancerous. Such growth cannot adapt to changing environmental factors or, if the growth survives, it becomes nonevolutionary and meaningless. As so succinctly stated by George Lock-Land in his *Live or Die:*

> . . . evolution carries with it the message of extinction. If we isolate ourselves from the system of life; if we do not make new friends of old enemies; if we do not find a balance of trade with our environment, we face a future known in all of its frightening dimensions. We, as a culture, as a species even, will join with the other unfit systems of our world. The

message is abundantly clear: Grow or die; evolution or extinction.

There are still ways for human civilizations to evolve on Earth in a positive fashion. But civilizations "off-Earth" are now available to us. It is an alternative offering greater mutuality and complementarity in evolutionary growth. We would be expressing the ultimate "death wish" of a species if we rejected this alternative, given the projected finiteness of Earth resources and general carrying capacity for a population.

However, once having accepted the option to populate space, we must cultivate space civilizations in such a way that they do not become replicated civilizations of Earthkind. They must not become clone civilizations. They must not be permitted to evolve as "enclaves of the Earth in space," as Gerard O'Neill seems to encourage. They must become independently unique, but still complementary to their progenitors.

Grow or die!

CHAPTER 9

Space War: Is It Inevitable?

The role of the military is to protect civilian interests, and as civilians migrate beyond Earth, the military will follow as long as there is a perceived potential enemy. This type of perception will not disappear until some fundamental changes occur in civilizations.

"Despite wishful thinking to the contrary," asserts General Jacob E. Smart (Ret.) in the Fall 1974 issue of *Strategic Review*, "man is and promises to remain an aggressive, combative creature. We fear, we hate, we fight one another. . . . We have no choice but to prepare to defend ourselves against attack in whatever form. . . . Today and henceforth the United States must be prepared to defend itself against aggression *in* space and *from* space. We cannot surrender that 'high ground' without contest."

However, Admiral Daniel Murphy, Deputy Undersecretary for Defense Policy, does not believe that war is inevitable. He is currently the Pentagon Analyst for Presidential Review Memorandum 23. He does feel the military has a role in space

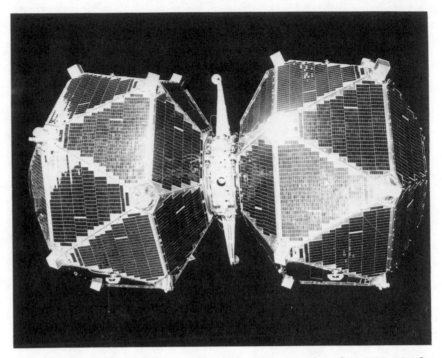

U.S. Air Force Vela satellites are designed to detect nuclear detonations in the atmosphere down to the earth's surface and over 100 million miles in outer space. *Courtesy of Space and Missile Systems Organization (SAMSO), U.S. Air Force*

for other purposes: "Satellites for verification of agreements and treaties is extraordinarily useful in preventing war."

Admiral Murphy goes on to say: "We would like to keep weapons out of space. President Carter is negotiating a treaty to prevent ASAT's (hunter-killer satellites) with the Soviet Union." Although he does not see space war as inevitable, he does not see its avoidance as inevitable either. "Since human beings are less than perfect, some policing in space is necessary *unless* it can be shown that there is something about space that will make us more perfect. Maybe if there were a way to cull out those who create conflict."

Others believe human nature continually improves, that technology has brought worldwide cultural sharing, religious values are becoming increasingly part of human behavior, and peace is a decision to be made by a free people. This point of view believes space war is preventable.

War in Space?

There is no consensus among the leading military, diplomatic, and think-tank analysts as to the causes of war, but there are some conditions with which it is associated. (See Simple Futures Wheel of Space War.)

First, there must be a sense of grievance about some action or actions that contradicted a moral standard or impinged on a vision of the future. Second, there must be two or more parties believing in an eventually successful outcome. Third, there has to be some objective of victory—the aggressor wants to take or keep something—that is believed to be in tune with the nature of things, as well as a belief that its intelligence and weapons are better than the other's. And fourth, there have to be weapons. In a "space war," the weapons could be frequency manipulation, particle accelerators, lasers, psychic or parapsychological "sorcerers," orbital rockets, and hunter-killer satellites (ASAT's).

An impending problem with military implications in space is the unilateral declaration by the Soviet Union that only satellite photos with resolution within 50 yards, or 150 feet, be considered military. In the meantime, the newest LANDSAT is broadcasting to Italy live photos of Europe—including parts of the Soviet Union—of 40 meters, or 131 feet.

Frequency manipulation devices in satellites or on the ground could cause electronic interference, false signals, and other forms of interference. Particle accelerators could send protron or electron beams to directly attack objects or could create and send plasmas or a super-gas state of matter around the world on the Earth's magnetic lines to blast areas like a gigantic lightning bolt.

A seemingly bizarre form of weaponry is the parapsychological "sorcerer." But they are being taken very seriously, and necessarily so.

John L. Wilhelm wrote an article called "Psychic Spying? The CIA, the Pentagon and the Russians Probe the Military Potential of Parapsychology" for the *Washington Post*, August 7, 1977, in which he stated that the ethics of parapsychology and its military implications are being studied around the world. The Stanford Research Institute is one such place; it has

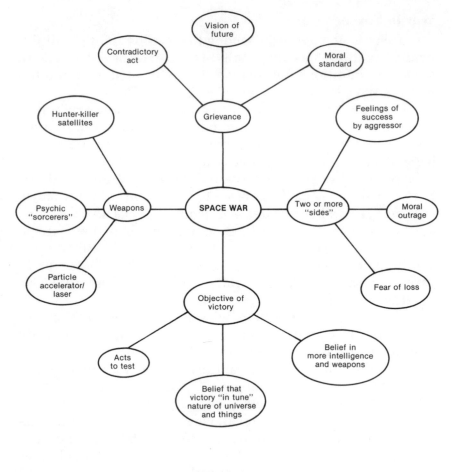

**Simple Futures Wheel
of Space War**

produced a report for the National Science Foundation entitled *The Use and Misuse of Consciousness Technologies.* Robert Toth, a *Los Angeles Times* reporter held by the Soviet KGB for possession of state secrets about parapsychology, is seriously mystified as to "why the Russians would classify parapsychological research as a state secret."

The Defense Intelligence Agency has detailed Soviet and Czechoslovakian experiments in the areas of parapsychology in a report called *Controlled Offensive Behavior—USSR (U).* The document claims that Soviet research could lead to: "1)

Knowing the contents of top secret U.S. documents of our troops and ships and the location and nature of our military installations; 2) Cause the instant death of any U.S. official at a distance; and 3) Disable, at a distance, U.S. military equipment of all types including spacecraft." The report goes so far as to claim that Soviet spies could be leaving their bodies to visit military installations—by 1979.

It seems like Carlos Castaneda's tales of sorcery are getting serious attention.

Who Wants War in Space?

The current and short-term space-war fighters are most likely the Soviet Union and the United States. Many Russians believe that America, in the midst of capitalism's "death throes," will resort to thermonuclear war. Hence the Soviet Union's large expenditure on domestic civil defense and their concomitant belief that they can survive the war even though they are not sure they can win. Such beliefs have motivated the steady growth of Soviet strategic forces over the years.

Americans, on the other hand, believe that thermonuclear war would be so terrible and destructive that it would mean the end of what we call civilization. Our policy has been called Mutual Assured Destruction (MAD) or, "If you try it, we all go down together." As a result, we don't take civil defense as seriously. (Compare the $1 to 3 billion dollars for Soviet civil defense to America's $64 million in fiscal year 1977.)

Will Soviet-American relations move from SALT to disarmament? Not according to one expression of Soviet policy in the November 1975 issue of *Communist of the Armed Forces:*

> The premise of Marxism-Leninism on war as a continuation of policy by military means remains true in an atmosphere of fundamental changes in military matters. The attempt of certain bourgeois ideologies to prove that nuclear missile weapons lead war outside the framework of policy and that nuclear war moves beyond the control of policy, ceases to be an instrument of policy, and does not constitute its continuation is theoretically incorrect and politically reactionary. . . .
> The description of the correlation between war and policy is fully valid for the use of weapons of mass destruction. Far

from leading to a lessening of the role of policy in waging war, the tremendous might of the means of destruction leads to the raising of that role. After all, immeasurably more effective means of struggle are now at the direct disposal of state power.

Space war, unfortunately, is far from out of the question.

U.S.-U.S.S.R. Space Arsenals

It is generally conceded that over half of all Soviet research and development is defense-related; that's three times our own. Twelve percent of the Soviet Gross National Product is devoted to defense, while we spend only 5 percent of our GNP on defense. The Soviet successful space launches have grown three-to-one over the United States in recent years (see chart).

Of these launchings, 326 were American military while 586 were Soviet military. The difference is even greater for military payloads, as the next chart indicates.

Artist's conception of one of twenty-four satellites that will make up NAVSTAR, a global navigation system for all U.S. military services. *Courtesy of U.S. Air Force Space and Missile Systems Organization (SAMSO)*

SUCCESSFUL MILITARY PAYLOADS
TOTALS 1957–1975 FOR U.S.S.R.-U.S.*

MILITARY PAYLOAD	U.S.S.R.	U.S.
Military Recoverable Observation	328	220
Minor Military (Environmental Monitoring, Radar Calibration, Electronic Ferret)	94	90
ELINT, FERRET (Satellite Picks Up Electronic Signal, Communications, and Radar Intelligence)	42	0
Navigation and Geodesy	46	31
Military Communications, Store-Dump	128	0
Early Warning Satellites (Senses and Transmits Electromagnetic Signals of Nuclear Explosions or Missile Launchings)	7	33
Fractional Orbit Bombardment System (FOBS)	18	0
Ocean Surveillance	12	0
Inspector Targets	9	0
Inspector Destructors (Hunter-Killer Anti-Satellite Satellites)	7	0
Orbital Launching Platforms	135	0
TOTAL	826	374

(Includes 48 DOD civilian payloads)

Source: Staff Report, Committee on Aeronautical and Space Sciences, United States Senate, by Charles S. Sheldon II, CRS Library of Congress

*Both the National Security Council and the Pentagon note there is a mistake in this table, but they are unable to supply clarification.

DSCS III, Defense Satellite Communications System satellite, is the newest and most sophisticated command, control and communications satellite, with narrow beams to Earth, increased antijam protection, and some defense against Soviet ASATS. The first two DSCS III's are scheduled for launch in 1979 and 1980. *Courtesy of SAMSO*

The United States has taken the "lean but lethal" approach, claiming that we still have superiority. The U.S. considered the development of both the hunter-killer satellites and Fractional Orbit Bombardment System (FOBS), but dropped them as not cost-effective. Instead, advanced laser systems will be used to counter these weapons. Barry Smernoff of the Hudson Institute feels that Soviet FOBS and hunter-killer satellites are decoys to force the U.S. to spend vast quantities of money. The U.S. military space budget for 1978 is $2.77 billion.

FOBS—in effect, part satellite, part ICBM—could fly a partial orbit around the earth to strike at an angle, reducing warning time from fifteen to six minutes. Although there have been at least eighteen Soviet FOBS flights, the United States does not consider those flights a violation of the space treaty that bans weapons of mass destruction from orbit, because so far they have flown less than one orbit and haven't carried nuclear warheads during testing and development. Yet the

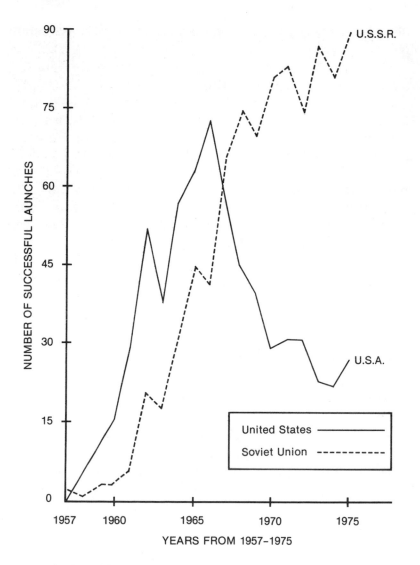

U.S.-SOVIET RECORD OF KNOWN SUCCESSFUL
SPACE LAUNCHINGS

Source: Staff Report, Committee on
Aeronautical and Space Sciences,
United States Senate,
by Charles S. Sheldon II,
CRS, Library of Congress

Chart by Future Options Room
Washington, D.C.

SUCCESSFUL LAUNCHES		
	U.S.	U.S.S.R.
Civil	323	292
Military	326	586

FOBS flights clearly violate the intent of the treaty; that seems obvious enough.

If Soviet FOBS had multiple maneuvering re-entry warheads, defense could be difficult. Apparently the U.S. believes the accuracy of our BMEWS (Ballistic Missile Early Warning System with fanned-out radar), our early warning satellites with infrared sensors, and our second-strike capability are still so devastating as to deter this use of the FOBS.

The Soviet hunter-killer satellites, tested at least eleven times, could certainly blind or destroy U.S. satellites. But these could be countered by high-powered lasers from Earth, from the satellite itself, or from the space shuttle. One can imagine some futuristic satellite/shuttle dogfights.

Both *Aviation Week* and *Newsweek* claim the Soviets

This is what the Soviet Hunter-Killer Satellite or Anti-Satellite Interceptor (ASAT) might look like: approximately three feet wide and ten feet long; infrared sensors are integrated with a small radar dish to seek out the target, most likely a low-orbiting reconnaissance satellite like the ten-ton "Big Bird." The ASAT is used to move an explosive device close to the target where it will explode with penetrating shrapnel. In September 1977, the U.S. awarded a contract to Vought Corporation to develop by the mid-1980s a Hunter-Killer satellite which, with on-board lasers, would vaporize target metal. Satellites are being hardened, i.e., made safer from an ASAT blast. *Courtesy of George Robinson*

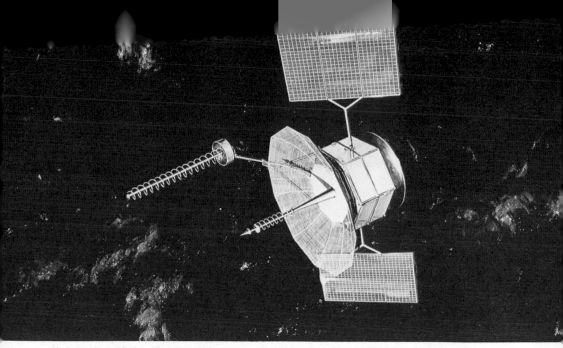

Artist's view of Fleet Satellite Communications (FLTSATCOM) System in equatorial orbit, vulnerable to ASAT and laser attacks. These satellites coordinate information for the U.S. Navy, Air Force, and the Department of Defense. *Courtesy of SAMSO*

were suspected of using laser weapons five times to temporarily blind the U.S. early warning satellites. Former Secretary of Defense Donald Rumsfeld said the malfunction was caused by intense light emitted by explosions in major gas pipelines in the western part of the Soviet Union. But there has been speculation that this explanation was manufactured to prevent the need for American retaliation; that would have destroyed then Secretary of State Henry Kissinger's attempt to achieve détente and would have hindered SALT negotiations as well.

According to *Aviation Week* of December 8, 1975, and as summarized by Charles Sheldon of the Congressional Research Service, "The United States, since the early 1960s, has probed Soviet satellites with lasers from Maui, Hawaii, and Cloudcroft, New Mexico, to determine lens and film types used in Soviet photographic missions, but not in a manner to cause deliberate damage to such Soviet satellites." It is not difficult to imagine a slow but steady escalation of check-countercheck leading to what some call the bloodless electronics war in space.

But such a limited space war seems unlikely. If it were to happen in the next fifteen to twenty-five years, there may well

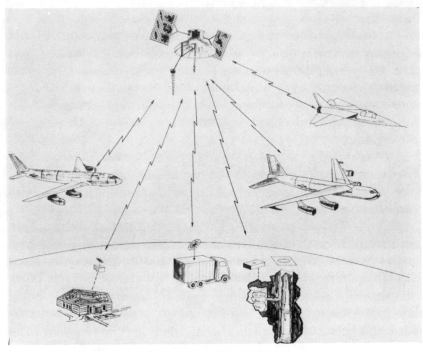

Courtesy of SAMSO

be large numbers of people in lunar and orbital space communities by that time. People would inevitably be involved in such a conflict, even if only as hostages. A space war between the United States and the Soviet Union is unlikely to be limited for two other very important reasons as well: (1) neither wants to be officially labeled a second-string power after a loss, and (2) space is the location of critical communications, navigation, and intelligence-gathering components of Earth-based weapons systems. As either country began to lose in space, the dependent ground, air, and sea weapons systems would be brought into play, escalating the warfare immeasurably.

Satisfied with the performance of FOBS and eleven orbital testings of hunter-killer satellites, the Soviets are now expanding their photographic reconnaissance space flights (see "Military Recoverable Observation" in the Military Payloads chart). They were used in greatest numbers during the 1973

Yom Kippur war—Kosmos 596, 597, 598, 599, 600, 602, 603.
Were these used to aid Egyptian command and control? Or for
Monday morning quarterbacking? Kosmos satellites 607, 609,
612, 616, and 625 were used to monitor the cease-fire agreement between Egypt and Israel after November 11, 1973. A
more sophisticated reconnaissance satellite than the Kosmos
fell out of orbit into Canada January 24, 1978—with a nuclear
power source.

Meanwhile, the American space shuttle adds a new
weapons delivery system to the possibility of space war. It can
place tens of satellites in orbit per single launch, thus eventually replacing the traditional rockets. It could inspect and
retrieve offensive weapons as well as command, control, and
communication systems in space. In fact, it could pick up an
entire Soviet Military Salyut (the small Soviet version of
Skylab) and return it to Earth. Even though the Soviets could
counter this with a simple self-destruct device, this nevertheless gives the Americans another pawn in a space war that the
Soviets so far lack.

One drawback to the U.S. Department of Defense contract-
ing shuttle flights to NASA is that it introduces the possibility
of conflicting—and time-consuming—interests within our

United Kingdom's Skynet Communications Satellite. *Courtesy of SAMSO*

NATO Phase III communications satellite for all NATO nations is integrated with the American DSCS III and British Skynet. *Courtesy of SAMSO*

own country. Since there are currently only four shuttles being developed, one can foresee military and civilian interests competing for their use. Interestingly, the March 3, 1975, issue of *Aviation Week* claimed Soviet scientists have approached NASA about joint shuttle flights.

So far, it seems that any space war would be confined to Earth orbit. But if breakthroughs in ion propulsion, fusion or gaseous core nuclear reactors occur, travel time throughout the entire solar system will be greatly reduced.

Futuristic Additions

Along with the space shuttle, high-energy lasers are considered among what Dr. Malcolm R. Currie, former Director of U.S. Defense Research and Engineering, called "Technological Directions of Great Promise" in his January 18, 1977, statement before the 95th Congress. Regarding the Soviets, he said, "There was an increase in the size of Soviet facilities that we know to be engaged in high-energy laser research and development from 1971 to 1975, and there are indicators which point to Soviet interests in particle beam technology which may have advanced weapon applications."

James W. Canan, in his book *The Super-Warriors*, said the CIA estimated the Soviets spent one billion dollars on high-energy laser research and development in 1974 alone. Knowing that, perhaps we can anticipate attacks made at the speed of light, a fundamental alteration of the traditional view of the bombers and missiles of nuclear war.

Not so, says Richard Garwin, IBM Research Fellow at T.J. Watson Research Labs. He feels the very development of high-energy laser systems is intolerable; the launching of such systems into orbit should be considered an act of war. But the same thing could have been said of the development of chemical warfare (in which the Soviets have a clear lead), ICBM's with nuclear warheads, or the hydrogen bomb.

Barry Smernoff responded to Garwin: "Even if it were established that the adversary's satellites contained HEL's [High Energy Lasers], there could remain the possibility that their purpose was to perform one of the *nonlethal* military applications of space-based HEL systems such as high-resolution imaging for space object identification, beamed power transmission, or intraspace propulsion." Could we discriminate among these uses accurately enough to attack appropriately, as Garwin suggests?

The point is that there are so many applications of lasers in space that their eventual development is inevitable, but their lethal military missions are not—if we work out "future HEL-oriented arms control agreements." These agreements should include the requirement that each party give prior notice to the other before electronically examining a satellite, since such actions could be misinterpreted as an attempt to cause damage and give rise to a chain reaction resulting in war. For that matter, prior notice to the U.N. of any space activity might be considered—and become—the best way to keep to the spirit of the U.N. Outer Space Treaty.

Smernoff argues that it is inevitable and desirable to re-place nuclear defense with laser and/or particle (electron or proton) beams. The Defense Intelligence Agency has already expressed concern about the potential Soviet use of micro-waves to alter behavior. Gerald Feinberg has theorized that there is a dimension beyond the speed of light; if so, lasers may in some future day be replaced by tackyon beams traveling faster than the speed of light. So far no one has found a tackyon, but if or when we do, Dr. Feinberg of Columbia University feels their first use would be in high-speed propulsion. It requires vast amounts of fuel to even begin to go a fraction of the speed of light. But with tackyons he feels the amount of fuel would be greatly cut. But he cautions that this will not by a long shot put us into *Star Wars'* hyperspace.

Another far-out possibility is the simulation of black holes. Could they, too, become a weapon in some future space war?

"Is there no end to this madness?" asks the robot C3PO in *Star Wars.*

Maybe Senators Jennings Randolph and Mark Hatfield have the right idea. They introduced a bill (S469) to Congress for a study of the establishment of a National Academy of Peace and Conflict Resolution. It passed the Senate June 17, 1977, and is now in the House (H.R. 8356). Such an academy could prove more than useful if it develops new approaches to joint U.S.-U.S.S.R. ventures, economics, and institutional arrangements. It could play a key role in space war prevention if it approaches the problem from a species and extraterrestrial view rather than simply within the antagonistic limitations of terrestrial nationalism.

Space War Prevention

The Americans and Soviets are engaged in an historic joint venture, the Strategic Arms Limitation Talks (SALT). It is quite unprecedented for two potential world power enemies to swap information about their weapons systems, and begin to set limits. Even though the Washington political satirist Mark Russell has called SALT, along with détente, "like going to a wife-swapping party and coming home alone," and even though SALT will certainly have its ups and downs, it marks a new chapter in civilization. Currently it is the only game in town that could even begin the process of disarmament.

In a paper entitled "Arms Control and Strategic Stability" presented by Sidney D. Drell at the National Academy of Sciences Annual Meeting on April 27, 1977, six criteria were listed as crucial to the Strategic Arms Limitation Talks:

(1) Clear evidence that neither country seeks strategic superiority.
(2) A rough overall equivalence of strategic forces.
(3) A mix of strategic forces designed so that a large fraction of each will survive a preemptive attack.
(4) Secure and effective command and control of one's nuclear forces.
(5) Maintenance of strong conventional non-nuclear military forces.
(6) The existence of "flexible response" alternatives to "massive retaliation."

Donald Brennan of the Hudson Institute has proposed the following ten-year agenda for SALT:

SAMPLE LIMITATION SCHEME FOR STRATEGIC
OFFENSIVE FORCES (Tons of Payload)

Category	Current U.S.	Current U.S.S.R.	Initial Ceiling	5-Year Ceiling	10-Year Ceiling
ICBM	1,000	3,000	3,000	1,500	750
M/IRBM	0	1,000	1,000	500	250
SLBM	1,000	880	1,000	500	250
Cruise Msls	0	1,000	1,000	750	500
Hvy Bmbrs	12,000	3,000	12,000	6,000	3,000
Med Bmbrs	2,000	6,000	6,000	3,000	1,500
NucTacAir*	X	Y	(X+Y)	(X+Y)/2	(X + Y)/4

The initial ceilings are set high enough to encompass the larger existing force in each category. Thereafter, the ceilings are decreased over time.

Note that a constraint in terms of payload does not imply a limit of one missile per ton. There could be more or fewer. (It would be preferable to have only one RV per ICBM.)

*Nuclear Tactical Aircraft. "X" stands for all American planes that can carry nuclear weapons. "Y" stands for Soviet planes. The exact number cannot be established now because conventional planes can be rewired.—Au.

It is extraordinarily difficult to build enough trust for systematic disarmament. It might become more possible to achieve cooperation if joint building efforts were conceived during dismantling. For example, if by 1980 the weapon systems were cut in half, as Donald Brennan suggests, then a building program using the disarmed nuclear warheads might help stimulate even further reduction. Such disarmed warheads could supply the fuel for a joint U.S.-U.S.S.R. space industrialization power supply or intersolar system space ship. Dr. Theodore Taylor, senior scientist of Project Orion, has designed a space vehicle that runs on nuclear fuel.

R. Buckminster Fuller proposed an electric power grid over the North Pole connecting the U.S., U.S.S.R., and China as a method to use power more efficiently. Local power grids are inefficient—they are not at peak use during the night. Such an international grid would optimize resources around the clock, around the world. Such a joint effort might be desirable enough to all concerned and help speed disarmament.

Dr. Gerard O'Neill believes some disarmed ICBM's and military boosters such as the Titan III-C could be used for launching joint U.S.-U.S.S.R. payloads into orbit, a sort of beating of swords into outer space plowshares.

The question for SALT is not how to negotiate a settlement of an unwanted technological fly in the peace ointment. Rather, it is how to use that fly productively as a part of the ointment. SALT is an unprecedented opportunity to turn a lemon into lemonade.

Other Hopeful Signs

The philosophies of the East and the West are merging from their historically separate courses. A reconciliation between the mystic and the technocrat is evolving. New Age human potential groups such as EST (Erhart Seminars Training), Lifespring, Arica, TM (Transcendental Meditation), and the like, are appealing to many different kinds of people, including the "technocrat." The peoples traditionally more mystically inclined than the rest of the world—primarily in the East—are trying technocratic management philosophy and satellite technology. The Soviet Union is paying increasing attention to the study of possible new perceptions of reality; this was documented in *Psychic Discoveries Behind the Iron Curtain*, by Lynn Schroeder and Sheila Ostrander. The world seems to be evolving a view of reality that might be referred to as "conscious technology." Put a bit more challengingly, "con-

Titan III-C develops 2.5 million pounds of thrust at liftoff, the largest American military space booster. *Courtesy of SAMSO*

sciousness creating technology creating consciousness, and technology creating consciousness creating technology."

Millions of people around the world are convinced that the universe is alive with extraterrestrial intelligence. President Carter has asked NASA to consider reopening UFO studies. It is a common assumption in both military and civilian circles that UFO landings would unite the human world as no internal stimulus could. The mere belief that we are not alone in the universe serves to create a feeling of kinship among humans.

Another facet of the emerging new perceptions of reality is that future Spacekind should be free from, and independent of, the political bonds of Earthkind. This view, explained in a later chapter, is a critical ingredient for the prevention of colonial space war.

The Block 5D Integrated Spacecraft System (ISS) gathers weather intelligence anywhere in the world. Its data is available through the U.S. Commerce Department's National Oceanic and Atmospheric Administration (NOAA). *Courtesy of SAMSO*

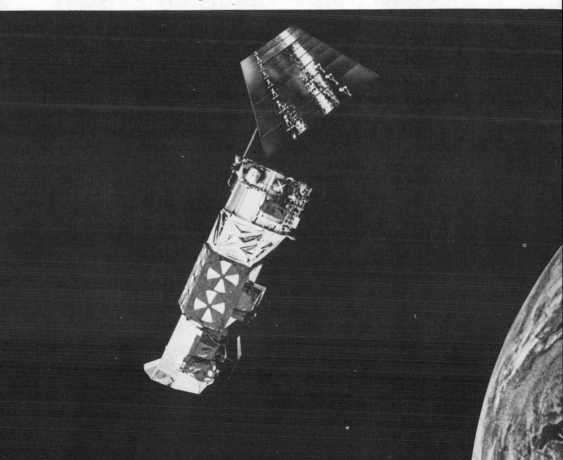

In the past, colonies always struggled to break free from the motherland or fatherland. We can expect the same struggle to occur between Spacekind and Earthkind, unless we agree now that future space citizens will be free of the earthly politico/cultural umbilicus. Granted, the early activities of the 1980s and 1990s will require the initiative and sustained support of Earth. But once these space communities establish themselves and pay back terrestrial investment, they should be granted independence—if they choose. The independence of Spacekind from Earthkind can reduce the likelihood of military conflict in space.

Furthermore, both capitalism and communism are beginning to be seen as early Industrial Age economic theories that are inadequate for the future. Some imaginative economists and Third World leaders realize that the old conflicts might be resolved if an entirely new economic theory could be developed.

It is a popular view that military expenditure and preparation for war are the most efficient way to create full employment. Guns put butter on the table. But this assumes that industry must be the basis of economic security. It is reasonable to assume that there will be future economies beyond industrialization that will have information and services as their main bases.

Perhaps the time is now to start developing the entirely new economic theory that can help resolve the conflict of communism and capitalism.

The growing power of the multinational corporation and the media is legendary. It has seriously eroded the power of the nation-states. The United Nations recognized the role of nongovernmental organizations in its Habitat Conference in 1976. Increasingly, politicians and theoreticians are beginning to realize that these world actors can play their roles peacefully if new relationships among them can be invented.

Rather than saying multinational corporations are good or bad, it is more useful to ask in what way they should relate to other world actors. Currently, the relationship between multinational corporations and nation-states is a bit fuzzy. The recent Lockheed briberies uncovered in Japan and the corporate

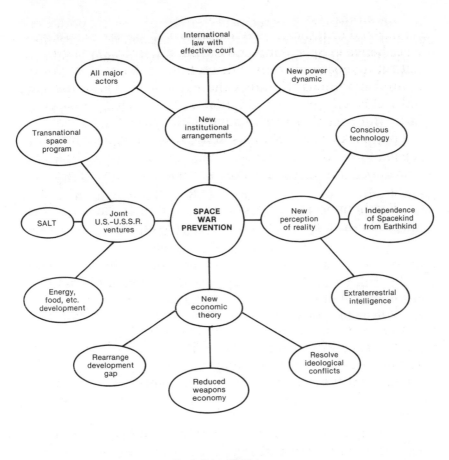

**Simple Future Wheel
of Space War Prevention**

embargoes of Israel leave government officials unsure of how to proceed.

Eventually, it will become apparent that the United Nations will have to create a new kind of world court—with an effective sanctioning capability—composed of members from the nation-states, multinational corporations, international associations, and the world media. If a nation felt a multinational corporation was trying to induce internal ferment to bring down prices of some needed commodity, that state could bring the grievance to the new world court. If the court ruled in favor of the state, then it might rule that the United Nations members place an additional 5 percent tax on all of that corporation's products as "punishment." Such sanctions could be very effective in restraining destructive behavior performed solely for power or profit.

We do not know just what new institutional arrangements will actually evolve in the future, but it is clear that these kinds of conversations will lessen the likelihood of war.

It is the job of advanced military scientists and engineers to come up with weapons systems that get around or beat anything treaties or agreements mention. The way to beat the arms race is with a new game. Joint U.S.-U.S.S.R. ventures, new perceptions of reality, the next economic theories, and new institutional arrangements are reasons to believe that space war could be prevented. A rapidly expanded space effort is the alternative to war; it will spur science and technology while creating full employment. But space development is not to be supported if it only means taking the same old cowboys-and-Indians war games *out there*.

CHAPTER 10

A Unifying Image
of the Future

Throughout history, civilizations have been held together by shared myths and common images of the future. When one civilization is to be bound with another in a peaceful way, joint myths must be created if they are to stay together. About 4,700 years ago when Upper and Lower Egypt were merged, the new unifying myth was that Ptah was the god of all creation and his temple at Memphis was to be the capital for the merged countries. In more recent time, Manifest Destiny bound together the peoples and states of Spanish origin in California with the French in New Orleans and the English pioneers moving across the Plains.

As the nation-states began solidifying their power and borders around the turn of the century, there was no common myth or image of the future. Some ethnocentric Europeans thought Christianity could provide the common vision, but it was fractionalized and in disarray across Europe. World War I offered the first stimulus to creating a world secular myth, since it was a world secular problem. Woodrow Wilson and others in

Europe pressed for the new myth—world peace through world institutions and law. The League of Nations was to be the confirmation that we had entered a new world order in the history of civilization. World War I was seen as "the war to end *all* wars."

Who could fault peace as the common image of the future to unite all the nations in cooperation? But it was not to be so. Wilson's plea to the American public was killed by the Republican Party. And World War II erased much of the "peace through world order" ideal, even though it did stimulate the United Nations, a far more effective world institution than the League of Nations.

But a new factor had entered the world dialogue for international cooperation: the atomic bomb. Very subtly the United States could say, cooperate or else! But others liked the atomic idea and soon the Soviet Union, along with England and more recently France, China, and India, have joined the nuclear club. Thus was generated the current myth: mutual cooperation with a balance of power, or mutual destruction. The U.S. policy, called Mutual Assured Destruction, abbreviates as MAD—a policy without a good fall-back position, as we learned in the Cuban Missile Crisis.

Searching for a Workable Image

So what *do* we need? A new vision of the future must be compelling, exciting to the imagination, and easy to sell. It must be multifaceted, pluralistic, and relativistic in order to appeal to the broad diversity of world cultures. And it must be inherently positive.

"Peace and love" seem to be too difficult to sell. Religions have pushed them for centuries. The fact remains that noble efforts to sell peace, such as the Kellogg-Briand Pact to outlaw war, signed in 1928 by the largest number of nations of any international agreement to that date, did not prevent World War II. There is no reason to believe that similar pacts and appeals to the humanity of people will succeed in the future to prevent World War III.

Such spiritual and intellectual efforts seek to change the

values of people and ignore the biologically based need for aggression. True, advances in agriculture and industry have greatly helped to satisfy basic human needs, making aggression itself less necessary. However, we are not out of the woods yet.

In the broadest sense, aggression is the power to act. A peaceful future with a no-growth society is contraindicated by aggressive needs. The thwarting of these tendencies may well lead to a violent expression of bottled-up aggressive needs.

Jane Roberts in *The Nature of Inner Reality* claims that aggressiveness can be a way of letting others know that someone has transgressed, and therefore can be a method of preventing violence.

Any creative idea is aggressive. Aggression leads to action, creativity, and life. Prevention of its expression can lead to destruction, violence, or annihilation. Hence, any future image must account for healthy expressions of aggression.

Manifest Destiny was an interesting shared image that matched our aggressive nature, was multifaceted, required little selling, had wealth as its reinforcement and reward, was seen as positive to those who believed in it, and definitely excited the imagination. But it crushed the American Indian, raped the land, rationalized slavery, and brought on territorial war. The sparkling rivers, buffalo, Indians, and Africans were not members of the Manifest Destiny image-making team.

We should not repeat this mistake on the world level. Instead, an essential criterion for the process of generating the next vision of the future should be world participation. This is not as insurmountable a problem as may first appear. For the first time in known history, we have international media with worldwide communications technology systems. The United Nations began a ten-year program of world conferences during the 1970s to stimulate national policies in energy, environment, food, shelter, etc. This valuable groundwork has begun an international dialogue on the future from the point of view of each national government.

Complex issues can become world discussion topics with much give and take. For example, the Club of Rome added the concept of limits to growth to the world dialogue; within two years it was a dominating view of world intellectual thought.

Recently, they have admitted this concept was wrong, and world opinion is beginning to turn around again. Therefore, complex issues can become part of the world dialogue, with opposing points of view heard and dealt with reasonably.

In a similar manner, this book is part of an effort to add outer space to the coming world dialogue on the next image of the future.

A New World Understanding

A new world understanding will accept that human destiny is beyond the solar system—that is, if humans are to survive and grow into the indefinite future. Remember, the sun blows up in ten billion years, so it's quite clear—we either blow up with it, or it blows up with human descendants remaining alive somewhere beyond the solar system. Between now and ten billion years, we will need a rather large-scale outward-bound program. Why not begin now?

Biologically speaking, everyone is the synthesis of his parents' genes, and in turn their grandparents', great grandparents', and so forth, all the way back to the very beginnings of time and creation. Every person alive today is the result of a continuous unbroken line of successes throughout evolution. Not bad! The genetic information we pass on to our children has been tested under evolutionary combat conditions and approved for the right to survive. Once we understand the potential of our intellect and that we are the grand genetic synthesis plus our current experiences, then we might begin to understand that we are also an expression of the universe. Hence, we have every right and responsibility to survive as long as possible.

What would have to happen first for the world to come to some saving understanding? We'd have to decide that the survival of humanity was worth the effort. Surprisingly, that may take some effort. We pollute the air we breathe and the water we drink. We rape the land and kill each other. "So what's worth the saving?" some ask. "Maybe the universe is better off without us."

The beginnings of a new vision of the future were put together at the Second General Assembly of the World Future Society in June 1975. The Committee for the Future ran an options room to condense and synthesize the thoughts of 3,000 futurists from around the world. The report from that options room called for:

> ... a pluralistic society which allows for continually expanding frontiers, choices for growth of human potential, and a secure environment for the opportunity to express love of your own style without reducing the options of others. This requires a new social consensus, a new perception of reality that is relativistic rather than dualistic, that your gain is my gain, that your loss is my loss, that is inclusive of difference rather than exclusive, that asks how can we do both rather than pitting one choice against another, and that values simplicity, clarity, and holds charity and honesty as pragmatic and prudent. This requires a world-wide and real-time interactive information system with easy access to all individuals. This new media gives and receives information when you want it, where you want it, and in the style you want it. It is the key condition to nurture decentralization. It gives local access to world resources and it allows all people the opportunity to evaluate, monitor and model the dynamic relationship between population distribution and resources distribution. The mistakes, injustices, and change options become apparent to all. It is the little boy saying the emperor is naked, the emperor has no clothes.

> This requires anticipatory democracy and participation among diverse peoples. Those affected by decisions must be involved with the decision-makers *before* the options for the decision are identified. Examples of this requirement are the variety of state goal-setting projects such as Iowa 2000, Washington Tomorrow, and the processes to do them such as SYNCON, Delphi, Charrette, and Shared Participation.

> Public participation processes can become real if we expect it to be more intuitive than analytical, more normative than descriptive, more decentralized than centralized, and more organic than mechanical. They can also be successful by ensuring that analytic, descriptive, centralized research in-

forms the process. Descriptive and normative forecasting/
planning should be seen as partners rather than one over the
other.

Such a new world understanding sees humans as the cut-
ting edge of evolution. With our new mental capabilities en-
hanced by advanced technology, we shall continuously be able
to create new world understandings, both on Earth and in outer
space. Since there are many different expressions of humanity,
there can be many varieties of new worlds, eventually creating
an ecology of humans, environments, and machines as richly
heterogeneous and interesting as earthly nature itself . . . and
possibly more peacefully symbiotic.

A new world understanding will redefine what it is to be a
person. Since Darwin, industrial society has accepted that we
were fighters in a dog-eat-dog world. The recent but strong
trend in Western thought is the acceptance of the expanded
Eastern view of consciousness. Essentially, the East sees
earthly existence as only *one* interesting expression of our
lives, while the West sees earthly existence as either *the only*
expression of our lives or as the test for acceptance into one or
the other final resting place. In short, the Easterner has a more
crowded universe of possible expressions of conscious life than
the Westerner does.

A new world understanding would not try to say that the
Eastern or Western orientation is the correct one for the future.
It would create a new synthesis of the two. It would develop
and value the concept of "conscious technology"—technology
to raise and expand consciousness, and consciousness to make
and use technology more powerfully, effectively, and aesthetically.

Our "self" is infinitely diversible, as Don Juan taught Carlos Castaneda in *Tales of Power*—"infinity surrounds you"—or
as Jonas Salk agrees when he says that genetically we are all of
the past universe and have genetic potentials far beyond our
current understanding. As we view ourselves as conscious
technology in process of infinite potential expression, so too
must we view others. This will result in a new reverence and
respect among humans. Those in the human potential move-

ment are just as naive in their antitechnology as are the technologists who turn up their noses at parapsychology.

The Need for Risk in Peace

It is hoped by several international affairs authorities that the goals either of peace or balanced ecology will provide the common purpose that will unite the world and prevent World War III. This is hopelessly romantic, because the image of a peaceful and clean environment just doesn't excite the imagination enough for people to take risks. The goal is too fuzzy and the image is too boring. *It is a vision of maintenance rather than a vision of growth.*

Mere maintenance as a goal lends no challenge to the imagination to bring forth the best from people. Maintenance is the vision of the nearsighted.

Peace and a clean environment can result from a new world understanding, but these should not be held up as the primary objectives or means for peacefully uniting the world. Instead, let's get the international media to begin putting the brushes in our hands to paint a new and common vision of the future that is compelling, excites the imagination, becomes easier to sell as it is generated by the world dialogue, is multifaceted, pluralistic, relativistic, positive, and enforceable.

CHAPTER 11

Earthkind—Spacekind

What will be the difference between the third-generation child born in space—in, say, the year 2040—and a child born the same year on Earth? It is generally agreed that the adventurous, intelligent, aggressive, and curious will move to space. Possibly, then, the first generation born in space will come from what we might call an "avant-garde" gene pool of the human species. These children will grow up with unique psychological pressure, knowing that they are special, knowing that they are the first of their species born outside the womb planet, and knowing the whole world is watching their development.

The great historian Arnold Toynbee believed that the environment a people lived in influenced their entire cultural history. Varying strengths of gravity, no atmospheric distortion of sunlight, and separation from Earth are some new stimuli that will contribute to the unfolding of a new history for Spacekind as distinct from Earthkind.

The combination of their genetic endowment, total physical environment, and psychological factors will make the first

"Comet Rendezvous II" by Rick Sternback. *Courtesy of the Smithsonian Institution*

NASA's Neutral Buoyancy Simulator helps orient astronaut engineers to the alien environment in which they will construct space facilities and communities. *Courtesy of NASA*

generation born in space unique, even from their rather already unusual parents. As this generation grows up, it is likely that they will develop new norms for raising their children. And by the third generation, the distinctions between Earthkind and Spacekind may be moving toward a species parallel to *Homo sapiens* but different from it.

We must now begin to consider how to create the correct conditions for the peaceful evolution of a culture and species we cannot fully predict. Fortunately, we have over a decade to discuss, experiment, and find the most agreeable strategy for space migration. What are the biological and psychological factors for consideration?

Biological and Psychological Factors

NASA's studies of astronauts' behavior indicate that synthetic and alien life-support systems can effect changes in the value-forming processes and ultimate behavior patterns. These changes can make long-duration or permanent space inhabitants significantly different from those who remain Earthsitters. It seems likely that Spacekind will perceive, conceive, and integrate their reasoning powers differently than if they were functioning in an Earthly environment. The effects of the synthetic foods and abnormal energy fields of Earth's highly industrialized society are bad enough; these effects will be even more telling on the relatively small space society living in the confines of a totally synthetic environment.

Health stabilization for inhabitants of space communities will present some unique problems. A way of life may very well evolve which emphasizes the necessity of complying with preventive measures to reduce the spread of respiratory, intestinal, and other forms of communicable infections. Public health procedures and regulations may become quite complex, with

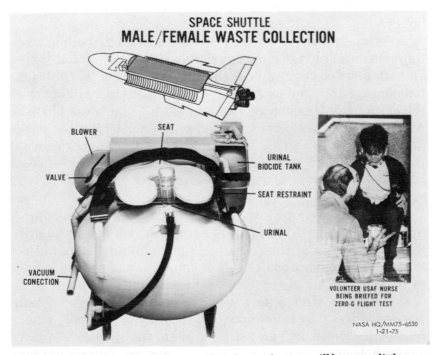

Those human functions we always considered second nature will become a little more complicated. *Courtesy of NASA*

serious criminal as well as civil penalties for violations. The problem of communicable disease is accentuated by the fact that the ongoing immunization process of mobile populations on Earth will not exist on the relatively immobile population of space communities. They will be more like remote and isolated tribes on Earth, susceptible to all the infections introduced by visitors from the outside.

Various experiments indicate that susceptibility to motion sickness, thresholds for perception of angular acceleration, perceived direction of internal and external space, refractive indices in the ambient atmosphere of a space habitat, and abnormal frequencies in the electromagnetic energy spectrum will cause variations in the manner in which space inhabitants touch, smell, see, perceive, conceptualize, and otherwise sense their surroundings and relationships. Since our bodies are highly complex, with interdependent systems, changes in one system affect other systems; even minor changes in the red blood cells, for example, affect the brain, endocrine system and bones.

Various experiments on Earth's surface carried out in environments designed to simulate that experienced by astronauts and other space community inhabitants indicate that substantial changes in social and work behavior patterns will occur in the actual environment of space habitats. And these changes will make it difficult for those who move between the space habitat and Earth's surface: they may be caught functioning on the earth's surface as though they were still in the space habitat and vice versa—kind of a super case of jet lag, culturally as well as physiologically.

Since the neurophysiological and psychological needs of Spacekind will be different from Earthkind, we must recognize these differences as we undertake the technological, architectural, and social design of the first generation of space communities. The impact of space environment realities on our traditional views of what a person or society will do under certain circumstances may well force us to reconsider precisely what a person is, could be, or perhaps even should be. We have suggested earlier that the catalyst for this new perception may be "conscious technology."

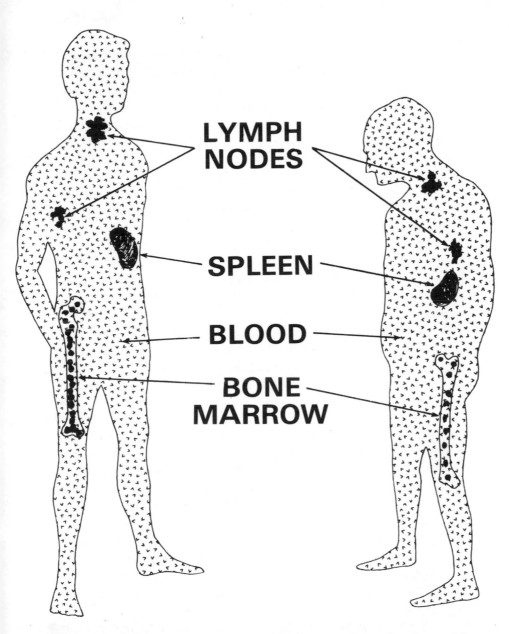

LYMPH NODES

SPLEEN

BLOOD

BONE MARROW

In animals, including man, the defenses against infection weaken with age as special immune cell groups in the lymph nodes, spleen, blood, and bone marrow produce fewer disease fighters. National Institute on Aging scientists are searching for the causes of age-related losses of immune response. Once the process is better understood, ways may be found to restimulate immune cells. This could lead to improved health for older people. *Courtesy of NIA Gerontology*

The hotly debated subject of recombinant gene research and the direct intervention in the essence of *Homo sapiens* through genetic surgery and positive eugenics (i.e., genetic counseling of couples who plan to have children, or who wish to know the odds for offspring with certain characteristics) could be very important in determining what the ultimate nature of Spacekind will be. We may have to rely on these technological capabilities to "engineer" individuals and societies which can survive in space.

But will all this selectivity and planning for space communities merely lead to a new, more sophisticated hedonism with advanced technological capabilities? Or will we have the wisdom to seek a well-integrated balance between our intense desire for physical-intellectual-spiritual evolution and a sensitivity to the natural environment, technological limitations, and economic demands?

It seems only reasonable that Spacekind's first generation or so will consist of individuals selected for their reliability and demonstrated professional competence, much like the present astronaut corps and space shuttle crews. The requirements for permanent space communities may demand a broader range of abilities, general knowledge, and flexibility in each individual, permitting a greater breadth of resources with which to make decisions in the daily operation and survival of a space community. Possibly a combination of the personality and mental types generally associated with Peace Corps volunteers and astronauts would make the most sense. Something as hard to define as wisdom may be a highly coveted and key characteristic of early space settlers. However, the art of manufacturing, even predicting, wisdom is not very precise at the moment.

What characterizes Spacekind? All of the diversity of Earthkind, altered only by a perception free of Earth's particular demands.

The "wisest available" was not necessarily the rule in the American colonies. Yet, it is generally believed that America's Founding Fathers were unusually wise. The survival skills of even first-generation space communities involve the mind and whole body far more than they did for the early American colonists. Will later groups be guided by the same capabilities

Artist's conception of future Spacekind living. *Courtesy of the Smithsonian Institution; Drawing by Sister Jessica, Foundation Faith*

and wisdom? Perhaps not. But the first social values and behavior patterns established will set precedents difficult to alter on the basis of later whim.

We must leave room in the occupation of outer space for different—perhaps conflicting—values and commitments. Even on Earth's surface we are fast learning that the ecology of diversity in species and behavior is far better for survival than inflexible and unaccommodating conformity and uniformity.

The planners must also view the communities as, in large part, experimental. Adaptations in behavior patterns will follow from new discoveries and inventions, unexpected stimuli and variations. Senses of unique identity will evolve. What may appear to be the irrational behavior of Spacekind must be accommodated by Earthkind within certain well-defined limits, so that the unique survival and cultural requirements of space society are protected.

With the new perceptions of morality emerging from the space environment, it is to be expected that unique values will evolve. Humankind tends to carry the old baggage of morality from one generation to the next. But morality is simply an intellectually codified set of responses accepted by civilizations as survival-oriented behavior patterns under a given set of influences. And these influences, or prevailing realities, change.

It has been asserted that ethics is a genetically dictated set of behavior patterns whose function it is to protect the continuous flow of a species' gene pool. They do not consist of a set of transcendant and immutable truths.

The only immutable and transcendant truth for *Homo sapiens* is the survival of ideas, knowledge, and wisdom. That is what gives meaning or purpose to physical or gene-pool survival, not abstract concepts of transient morality.

A sensitive change in that survival code appropriate to space existence should be recognized and accepted by both Spacekind and Earthkind. We will be in a new situation that will require new survival values. The old ground rules of society will be held in common—if only in memory—and provide the sense of continuity necessary for social evolution.

On the other hand, it is ridiculous to believe there won't be a revolutionary "morality" evolving in space communities. We shall have to create new social structures to handle the tensions between the old and new "morality" conflicts.

Recently there has been much discussion of the possibility of the establishment of a "science court" to resolve or synthesize the views of two or more scientific authorities in conflict. We might need a similar "court" to resolve the conflict between survival values in space with the traditional values of civilizations that have evolved in the equally unique environment of Earth's surface. Questions of genetic engineering, cloning, brain stimulation, and the like, will continually come up for decision and redecision.

It is a peculiarly American assumption that positive growth will occur financially, spiritually, and culturally simply by letting people rule their own destinies in their own fashions. If that is true, such a growth process will expand differences between Earthkind and Spacekind, possibly leading to serious conflicts. Even differences between lunar citizens and free-space citizens might well appear; lunar communities might tend to be more culturally conservative or Earth-like than those of free space.

Writer Clifton Fadiman sets out some questions, shared by many people who are strong proponents of space migrations and orbiting communities. "Just what kinds of men and women," he inquiries, "are most naturally conditionable for the space humanizing business? Should we not at least ask such questions before we colonize or industrialize or humanize space? . . . Technophobes like me have a haunting vision of a possible coming society composed mainly of two classes: master technocrats (and their military and bureaucratic colleagues) and happy, comfortable helots."

The rebuttal to Fadiman's haunting vision lies in the recognition and acceptance of the distinction between cultural values that have evolved on Earth and those which must be developed and protected for long-duration or permanent space communities. The "business as usual" attitude of many political scientists and lawyers involved with space activities might well serve the needs of existing space technology; but that attitude seems insensitive to the complicated and sensitive planning required for inhabitants of long-duration and permanent space communities.

We are beginning to realize that traditional economic theories and political ideologies may well provide unstable foundations for space community cultures. If we are to avoid perpetuating violent conflicts in the space arena, a broader interdisciplinary array of scientists, artists, lawyers, engineers, theologians, economists, futurists, etc., must work together to help meld new economic concepts with established and unique cultural realities that protect both Spacekind and Earthkind; that cultivate independent cultural needs and healthy economic symbiosis at the same time. Such a group should also be interested in giving—participatory planning is more easily accepted and easier to implement. Traditional economic, political, and ideological conflicts inherent in policies of "colonialism," for example, must be avoided in populating space.

Every legal, ideological, and technological safeguard possible must be used to protect Spacekind from the historically predictable aberrations of Earthkind's earthbound reason and logic. Space must be allowed to move to a higher or richer dimension of cultural evolution.

There is a parent-child relationship between Earthkind and Spacekind that must be recognized now in order to plan properly for the independent cultural evolution of Spacekind; a child grows up and strikes out on his own when he has sufficient maturity and resources. To accept less of a mandate is to lay the foundation for colonial conflict in the future.

Convention for Earthkind and Spacekind

The emphasis of scientists contemplating the cultural evolution of space communities seems to have been, so far, upon the need for independence from Earth cultures—the need to evolve a unique identity. That emphasis might appear a bit specious in the absence of a good grasp of the economic interdependencies that will necessarily either be established or evolve. However, if an international convention were to be formulated by Earthkind that would encourage the cultural independence of Spacekind, perhaps the basic principles would include the following:

1. The exploration and use of outer space, including all celestial bodies, should be carried out for the benefit and in the interests of not only Earthkind, but also of *Homo sapiens* migrants to outer space—that is, of Spacekind. Such areas of habitation should be considered the province of Spacekind first, Earthkind next.

2. Space communities and Spacekind, including those on or beneath the surface of other celestial bodies, should not be subject to national sovereignty or citizenship deriving from, or exercized by, nation-states or regional jurisdictions originating on Earth. Such communities should exercise independent cultural and political sovereignty which in no manner would relate to any territory or geographical boundaries.

3. Signatories of the convention should conduct their relationships with each other and with Spacekind in a manner consistent with international law, the Charter of the United Nations, or any successor organization, in the interest of maintaining peace and security and promoting cooperation and understanding between Earth cultures and the behavior patterns unique to space communities and civilizations.

4. Although the use of military personnel for scientific research or for any other peaceful purposes requiring interaction with outer space communities and Spacekind should not be prohibited, there should be no bilateral or regional military relationships or alliances whatsoever established between Earthkind and any space community or its inhabitants. Also, each nation should bear international responsibility for its own national activities in outer space that may directly affect outer space communities or their individual inhabitants.

5. Earthkind should pursue studies to ascertain what harmful effects might result from the introduction of alien material or the imposition of alien cultural characteristics on the ecosystems and cultural integrity of outer space communities.

6. In order to ensure the integrity of the peaceful purposes and intents embodied in the convention, each member should establish permanent space communities in such a manner that

they would be open for cultural examination and military investigation by representatives of other signatory states. Such examinations and investigations should not occur as a matter of right beyond the second generation born in any space community.

7. Members of the convention should agree to the establishment of an expert organization, under the aegis of the United Nations or its successor, which perhaps might be referred to as the International Organization for Sentient Space Activities (IOSSA). The principal purposes of this organization, established under separate charter, would be twofold: (1) to provide an international academy of behavioral scientists to review constantly all aspects of interactive relationships between Earthkind and Spacekind occurring either in outer space or on Earth's surface; and (2) to refer case situations to an appropriate international court of justice or arbitration whenever the propriety and predictable compatibility of such interactive relationships are at issue among experts.

EARTHKIND	SPACEKIND
Gravity only	Degrees of gravity from zero to more than 1g
Feeling average	Feeling special
Does not see the earth as a whole	Daily sees the earth as a whole
Both natural and built environment	Totally built environment
Different distribution of electromagnetic spectrum	
Trial and error evolution	Participatory planning
Atmospheric distortion of sunlight	No (or little) atmospheric distortion of sunlight

Psychological base: Earth	Psychological Base: Transition
Up/down perception of space	In/out perception of space
Partial control over temperature, humidity, seasons, length of daylight, weather, gravity, and atmospheric pressure	Total control over temperature, humidity, seasons, length of daylight, weather, gravity, and atmospheric pressure

CHAPTER 12

"Declaration
of Independence"

We do not assume all suggestions presented in this book for peaceful movement from Earth to space will be followed. It is more likely that Earthkind will try to exert control over Spacekind longer than is tolerable. It is more likely that the first permanent manufacturing facility will be sustained and operated on a colonial basis. If so, there will be very little practical recognition of the unique needs, social values, behavior patterns, and intellectual requirements of Spacekind.

But it will become apparent to the early space inhabitants that when a community becomes so effective that it can create a unique people and environment, then those people are by right free and independent from their original linkages. What might well emerge is a document like the following, adapted from the American Declaration of Independence.

In Representative Assembly of Space Migrants
The Unanimous Declaration
of the Communities of Space Migrants

Recognizing the distinction between thought processes occurring in space and those responding to the influences of inhabiting Earth's surface;

Believing that the habitation of space should be characterized by the full expression of the limitless varieties of human-related cultures;

Believing that an accurate understanding of the biological foundations of value-forming processes in a space environment will contribute substantially to lessening the fruitless competition and violent conflicts among Earth civilizations;

Desiring to elevate the evolution of *Homo sapiens* to its next logical stage,

Be it therefore DECLARED:

WHEN IN THE COURSE OF HUMAN EVOLUTION it becomes necessary for progeny to dissolve the political and biological bonds which have connected them with their progenitors, and to assume among the powers of the solar system and galaxy the separate and equal station to which the Laws of Nature and their Creator entitle them, a decent respect to the opinions of Earthkind requires that they should declare the causes which impel them to their separation into Spacekind. We hold these truths to be self-evident, that Earthkind and Spacekind are created equal to their own respective environments, that once having been raised above their biological origins to a recognizable level of sentience and sapience they are endowed by their Creator with certain inalienable rights, and that among these rights are survival, freedom of thought and expression, and the evolution of individual and community knowledge. That to secure these rights, governments are instituted among sentient beings, deriving their reasonable and responsive powers from the consent of the governed and by protective inference from those life forms without the capacity to communicate interspecies. That whenever any form of government becomes destructive of these ends, it is the right of the

Time-lapse photograph of our Milky Way Galaxy as viewed from Yucaipa, California in 1977. Turn the book clockwise one-quarter turn so that the silhouetted trees are at the lower right bottom of the page and the Galaxy runs vertically. Now you can see where you, the trees, and the Earth are in relation to the center of the Galaxy, which is the large bulge of stars just above the trees. The Galaxy is 100,000 light years in diameter and Earth is 25,000-30,000 light years from its center. There are estimated to be 250 billion stars in our Galaxy and 100 billion galaxies in our universe. *Photograph by Patrick S. Michaud. Courtesy Richard Hoagland*

governed to alter or abolish it, and to institute a new set of
values and political framework, laying its foundation on such
principles and organizing its duties and authority in such form
as to them shall seem most likely to effect both their physical
safety and sense of well-being through cultural evolution. Pru-
dence, indeed, will dictate that political, economic, and
ideological traditions long established should not be changed
for light and transient causes; and accordingly all experience
has shown that Earthkind, and now Spacekind, are more dis-
posed to suffer, while evils are sufferable, than to right themselves by abolishing or radically restructuring the forms to
which they are accustomed. But when a long train of abuses,
usurpations, and insensitivity to the needs of future generations evolving in a unique life support environment, pursuing invariably the policies of colonial dependency and biological parochialism, it is their right, their obligation, to destroy
such usurpations, insensitivity, and unresponsive institutions,
and to provide new value standards that will ensure their
security from abuses by progenitor cultures and governments
of Earthkind. Such has been the sufferance of space community
migrants who are now evolved to Spacekind, and who now of
necessity are constrained to alter the existing foundations of
relationships among Earthkind and Spacekind. The history of
governments and private enterprise in space development industries is a continuing history of injuries and usurpations, all
having in direct object the maintenance of an absolute tyranny
over space communities and Spacekind. To prove this, a list of
grievances is unnecessary. A candid world need only remind
itself of the historical patterns of Earthkind when nations have
pursued economic, ideological, and religious expansion into
the less technologically developed continents and societies of
Earth. The plea of this Declaration is to break the cyclic violence, warfare, and destruction of civilizations which follow
with certainty from the establishment of colonies. We have
petitioned for redress in the most humble terms: Our repeated
petitions have been answered only by repeated neglect. We
have warned the governments and appropriate controlling in-

terests of Earthkind from time to time of their determined insistence to extend their total jurisdiction over space communities and Spacekind functioning in an Earth-alien environment. We have reminded them of the circumstances of our emigration and settlement in space, and those of our predecessors. These warnings and reminders, too, have met with the deafness of prevailing justice and a failure to recognize the responsibilities of consanguinity in succeeding generations of Earthkind. We must, therefore, denounce the causes and acquiesce in the necessity of our separation, and hold them, as we hold the rest of Galactic intelligence, Enemies in war; in Peace, Friends.

We, therefore, the representatives of space migrants, space communities, and Spacekind descendants of Earthkind, appealing to the Creator for the rectitude of our intentions, do, in the name and by the authority of Spacekind settled and living in space communities, solemnly publish and declare that these communities and their inhabitants are free and independent; that they are absolved from all allegiance to the governments and organizations of Earth; and that all political and ideological subservience of Spacekind to Earthkind is and ought to be totally dissolved; and that as free and independent communities of Spacekind they have full power to protect themselves, establish peaceful relations, contract commercial and defensive alliances, and to do all other acts and things which independent communities in space, as well as on Earth, may do. And for the support of this Declaration, with a firm reliance on the protection offered through the Creative Intent, we mutually pledge to each other our lives, our fortunes, and our Sacred Honor.

Time Ahead

Humanity has ahead of it the opportunity for a future rich in ways undreamed of before in all its past. The concepts exist, the technology is developing, and the will is growing. The dream need not be a dream any longer.

The time is here for H.G. Wells's vision to become reality:
let us "laugh and reach [our] hands amidst the stars."
And let us start now.

Spaceflight Bibliography

This bibliography is based on "Long-Range Planning for Space Flight," compiled by the Office of Space Flight of the National Aeronautics and Space Administration. Several recent publications have been added.

GENERAL

Berry, Adrian. *The Next Ten-Thousand Years: A Vision of Man's Future in the Universe.* Saturday Review Press, 1974.

Cole, Dandridge M. *Beyond Tomorrow: The Next 50 Years in Space.* Amherst Press, 1965.

Ehricke, Krafft A. "A Long-Range Perspective and Some Fundamental Aspects of Interstellar Evolution." *Journal British Interplanetary Society* 28:713–734, 1975.

——"Extraterrestrial Imperative." *Bulletin of the Atomic Scientists* 27: 18–26, November 1971.

—— "In-Depth Exploration of the Solar System and Its Utilization for the Benefit of Earth." *N.Y. Academy of Sciences* 187:427 et seq., 1970.

————"A Strategic Approach to the Development of Geolunar Space." Paper SD69-710, IAA Orbiting International Laboratory and Space Sciences Conference, Cloudcroft, N.M., Oct. 1969.

———— "Toward a Three-Dimensional Civilization." *Space World*, December 1970.

Harkins, Arthur M. "Humanism in Space." *Futurics* 1, no. 3: 79–88, Winter 1976.

Hearth, Donald. "Outlook for Space." NASA SP-387, January 1976.

Kahn, Herman *et al. The Next Two Hundred Years*. Morrow, 1976.

Powers, Robert M. *Planetary Encounters: The Future of Unmanned Spaceflight*. Stackpole Books, 1978.

Sagan, Carl, and Shklovskii, I. *Intelligent Life in the Universe*. New York: Delta Books, 1966.

Sagan, Carl. *The Cosmic Connection*. New York: Dell Publishing Company, 1973.

Sauber, William J. *The Fourth Kingdom*. Midland, Michigan: Aquari Corp., 1975.

Von Puttkamer, Jesco. "Developing Space Occupancy: Perspectives of NASA Future Space Program Planning." *Space Manufacturing Facilities (Space Colonies)*, ed. Jerry Grey. AIAA publication, March 1, 1977.

————"On Man's Role in Space: A Review of Potential Utilitarian and Humanistic Benefits of Manned Space Flight." Monograph, Washington, D.C.: NASA Office of Space Flight, December 1974.

————"Reflections on a Crystal Ball: Some Thoughts on the Relevance of Science Fiction." *Futurics* 1, no. 4: 152, Spring 1977.

————, ed. "Space for Mankind's Benefit." Washington, D.C.: Special Publication SP-313, Government Printing Office, 1972.

Worden, Arthur M. *Hello Earth: Greetings from Endeavor*. Nash Pub., 1974.

SPACE INDUSTRIALIZATION

Adams, C. Mel. "Production, Assembly, and High Vacuum Fabrication." *Space Manufacturing Facilities (Space Colonies)*, ed. Jerry Grey. AIAA publication, March 1, 1977.

AIAA/NASA Symposium on Space Industrialization, NASA-Marshall Space Flight Center, Huntsville, Alabama, May 26–27, 1976.

Bekey, Ivan, and Mayer, H.L. "1980–2000: Raising Our Sights for Advanced Space Systems." *Astronautics & Aeronautics*, July/August 1976.

Bekey, Ivan; Mayer, H.L.; and Wolfe, M.G. "Advanced Space System Concepts and Their Orbital Support Needs (1980-2000)." Aerospace Corp. Report ATR-76 (7365)-1, vols. 1–4. Contract NASW 2727, April 1976.

Bekey, Ivan. "Potential Space System Contributions in the Next 25 Years." *Future Space Programs 1975*, vol. 2. U.S. Government Printing Office, September 1975.

Bredt, J.H., and Montgomery, B.O. "Materials Processing in Space— New Challenges for Industry." *Astronautics & Aeronautics* 13, May 1975.

Ehricke, Krafft A., and Newsom, B.D. "Utilization of Space Environment for Therapeutic Purposes." AAS 12th Meeting, San Diego, CA, February 1966. ASS Preprint 66-19.

Ehricke, Krafft A. "Cost Reductions in Energy Supply through Space Operations." Paper presented at 27th International Astronautical Congress, Anaheim, Calif., October 1976, Session 34, paper IAF-A-76-24.

——"Lunar Industries and Their Value for the Human Environment on Earth. *Acta Astronautica* 1: 585–682, 1974.

——"Space and Energy Sources. Space Division, Rockwell International Corp., May 1977.

——"Space Industrial Productivity, New Options for the Future." U.S. Congress, House of Representatives, Serial M, vol. 2, Government Printing Office, September 1975.

——"Space Tourism." ASS Preprint 67-127, Dallas, May 1967.

——"The United Nations and the Power Relay Satellite as Elements of Global Energy Development." Report KE75-2-1, Rockwell International Space Division, April 5, 1975.

Gilmer, J.R., ed. "Commercial Utilization of Space." *AAS Advances in the Aeronautical Sciences* 23, 1968.

Glaser, Peter E.; Maynard, Owen E.; Mackovciak, J.; and Ralph, E. "Feasibility Study of a Satellite Solar Power Station." NASA Contractor Report CR-2357, Washington, D.C., February 1974.

Glaser, Peter E. "Development of the Satellite Solar Power Station." *Space Facilities (Space Colonies)*, ed. Jerry Grey. AIAA publication, March 1, 1977.

——"Power from the Sun: Its Future." *Science* 162: 857–861, November 22, 1968.

——"The Satellite Solar Power Station: An Option for Energy Production on Earth." AIAA paper, 1965.

Lessing, L. "Why the Shuttle Makes Sense." *Fortune* 85: 93-97, 113, January 1972.

McDonnell Douglas Astronautics Company-East, "Feasibility Study of Commercial Space Manufacturing." MDC #1400, 1975.

Minney, O.H. "The Utilization and Engineering of an Orbital Hospital." AAS Meeting, Dallas, May 1967. AAS Preprint 67-123.

Patha, J.T., and Woodcock, G.R. "Feasibility of Large-Scale Orbital Solar/Thermal Power Generation. Proceedings, Intersociety Energy Conversion Engineering Conference, AIAA, New York, 312–319, 1973.

"Skylab Results." Proceedings of 3rd Space Processing Symposium, NASA-Marshall Space Flight Center, Huntsville, Alabama, April 30–May 1, 1974.

"Skylab Results." Proceedings of 20th AAS Annual Meeting, August 20–22, 1974.

Stine, G. Harry. *The Third Industrial Revolution.* G.P. Putnam's Sons, 1975.

U.S. Congress, House Committee on Science and Technology. "Space Shuttle 1975." Government Printing Office, 1975.

U.S. Congress, House Committee on Science and Astronautics. "Space Shuttle—Skylab: Manned Space Flight in the 1970's." Government Printing Office, 1972.

U.S. Congress, Senate Committee on Aeronautical and Space Sciences. "Solar Power from Satellites." Government Printing Office, 1976.

U.S. National Research Council, Space Applications Board. "Practical Applications of Space Systems: Materials Processing in Space." Government Printing Office, 1975.

von Puttkamer, Jesco. "The Next 25 Years: Industrialization of Space." *Journal of British Interplanetary Society,* July 1977.

————"World Workshop in Space—But Built to Human Scale: NASA Plans for the Next 25 Years." *The Engineer,* June 1977.

Williams, J.R. "Geosynchronous Satellite Solar Power." *Astronautics & Aeronautics* 13: 46–52, November 1975.

Woodcock, Gordon R., and Gregory, D.L. "Derivation of a Total Satellite Energy System." AIAA Paper 75–640, 1975.

Woodcock, Gordon R. "Closed Brayton Cycle Turbines for Satellite Solar Power Stations." *Space Manufacturing Facilities (Space Colonies),* ed. Jerry Grey. AIAA Publication, March 1, 1977.

SPACE STATIONS, SPACE BASES, AND TECHNOLOGY

AIAA Technical Activities Committee. "Earth-Orbiting Stations." *Astronautics & Aeronautics* 13: 22–29, September 1975.

Baker, David. "The Problem of Space Station Longevity." *Spaceflight* 16: 303–304. August 1974.

Blagonravov, Anatolij A. "Space Platforms: Why They Should be Built." *Space World* H-8-92, August 1971.

Bono, P., and Gatland, K.W. "The Commercial Space Station." *Frontiers of Space*. Macmillan Co., 1970.

Drake, G.L.; King, C.D.; Johnson, W.A.; and Zuraw, E.A. "Study of Life-Support Systems for Space Missions Exceeding One Year Duration." *The Closed Life-Support System*. NASA Ames Research Center, NASA SP-134, 1967.

"The Effects of Confinement on Long-Duration Manned Space Flight." Proceedings of NASA Symposium, NASA Office of Manned Space Flight, 1966.

Ehricke, Krafft A. "Beyond the First Space Stations." AIAA Meeting, NASA Marshall Space Flight Center, Huntsville, Alabama, January 1971.

———"Space Stations: Tools of New Growth in an Open World." Lecture, International Astronautical Federation congress, Amsterdam, Netherlands, October 1974.

Hagler, Thomas A. "Building Large Structures in Space." *Astronautics & Aeronautics* 14: 56–61, May 1976.

Kline, Richard L. "The Space Station and Space Industrialization." Paper presented at AAS/AIAA Bicentennial Space Symposium, Washington, D.C., October 6–8, 1976.

Low, George M. "Skylab . . . Man's Laboratory in Space." *Astronautics & Aeronautics* 9, no. 6, June 1971.

Marshall Space Flight Center. "Manned Orbital Facility: A User's Guide." Government Printing Office, 1975.

McDonnell Douglas Astronautics Co. "Manned Orbital Systems Concepts (MOSC) Study." Contract NAS8-31014, October 1975.

———"Space Station—Executive Summary." Contract NAS8-25140, August 1970.

———"Space Station Program Definition Study/Space Base." MSFC-DRL-140. Contract NAS8-25140, 1970.

———"Space Station Systems Analysis Study." Part 1 Final Report, MDCG6508, vols. 1–3, September 1976. Part 2 Final Report, MDC

G6715, vols. 1–3, February 1977. Contract NAS9-14958, NASA/JSC.

———"Space Station: User's Handbook." Report MDC G-0763, March 1971.

North American Rockwell, Inc. "Space Base." MSC-00721. Contract NAS9-9953, July 1970.

Parin, W. "Life on Orbital Stations." *J. Aerospace Medicine* 41, no. 12, December 1970.

Rockwell International Company. "Austere Modular Space Facility." SD 75-SA-0105, September 1975.

———"Modular Space Station Phase B Extension." Contract NAS9-9953, NASA/JSC.

———"Observation, Assembly, Staging and Industrial Support (OASIS) Facility." SD 75-SA-0106, September 1975.

———"Space Station Concepts." SD 760SA-0072.

———"Space Station Systems Analysis." SD 75-SA-9301, February 1976.

Smith, T.D., and Charhut, D.E. "Space Station Design and Operation." AIAA *J. Spacecraft and Rockets* 8, no. 6, June 1971.

LUNAR BASES AND SPACE COLONIZATION/SETTLEMENTS

Asimov, Isaac. "After Apollo, A Colony on the Moon." *The New York Times Magazine* 30, May 28, 1967.

———"Colonizing the Heavens." *Saturday Review* 17, June 28, 1975.

———"The Next Frontier?" *National Geographic Magazine*, 76–89, July 1976.

Brand, Stewart, ed. *Space Colonies.* Penguin Books, 1977.

Chedd, Graham. "Colonization at Lagrangia." *New Scientist*, October 24, 1974.

Chernow, Ron. "Colonies in Space." *Smithsonian*, 6–11, February 1976.

Clarke, Arthur C. *Islands in the Sky.* NAL Signet Book, 1965.

———*Rendezvous with Rama.* Harcourt, Brace, Jovanovich, 1975.

Cravens, Gwyneth. "The Garden of Feasibility." *Harper's Magazine*, August 1975.

Dempewolf, Richard. "Cities in the Sky." *Popular Mechanics*, 94–97, 205, May 1975.

DeWane, Michael. "Space Colonies—Environmental Boon or Bane?" *Futurics* 1, no. 2: 20–23, Fall 1976.

Dossey, J.R., and Trotti, G.L. "Counterpoint—A Lunar Colony." *Spaceflight* 17, July 1975.

Driggers, Gerald W., and Newman, J. "Establishment of a Space Manufacturing Facility." AIAA *Progress in Aeronautics and Astronautics* series, 1977.

Driggers, Gerald W. "A Factory Concept for Processing and Manufacturing with Lunar Materials." 3rd Princeton/AIAA/NASA Conference on Space Manufacturing Facilities, May 9–12, 1977. AIAA Paper 77-538.

————"Systems Analysis of a Potential Space Manufacturing Facility." 3rd Princeton/AIAA/NASA Conf. on Space Manufacturing Facilities, May 9–12, 1977. AIAA Paper 77-554.

Erickson, Scott W. "The High Frontier: Space Colonization and Human Values." *Futurics* 1, no. 2: 24–28, Fall 1976.

————"Space and the Exploring Spirit." *Futurics* 1, no. 4: 134–144, Spring 1977.

Friedman, Phillip. "Colonies in Space." *New Engineer*, November 1975.

Guillen, Michael. "Moon Mines, Space Factories and Colony L-5." *Science News*, August 21, 1976.

Heppenheimer, Thomas A., and Hopkins, Mark. "Initial Space Colonization: Concepts and R&D Aims." *Astronautics & Aeronautics*, 58–72, March 1976.

Heppenheimer, Thomas A., and Kaplan, D. "Guidance and Trajectory Considerations in Lunar Mass Transportation." *AIAA Journal* 15, no. 4, April 1977.

Heppenheimer, Thomas A. *Colonies in Space.* Stackpole Books, 1977.

————"Steps Toward Space Colonization, I. Achronmatic Trajectories." April 19, 1977, to be published.

————"Steps Toward Space Colonization, II., Colony Location and Transfer Trajectories." April 23, 1977, to be published.

————"Two New Propulsion Systems for Use in Space Colonization." *J. Brit. Interplanetary Soc.* 31, 1977.

Hopkins, Mark M. "Cost-Benefit Analysis of Space Manufacturing Facilities." 3rd Princeton/AIAA/NASA Conference on Space Manufacturing Facilities, May 9–12, 1977. AIAA Paper 77-554.

Johnson, Richard D., and Holbrow, C., eds. "Space Settlements—A Design Study." NASA Ames Research Center, NASA SP-413, 1977. Government Printing Office, Washington, D.C.

La Petra, J.W., and Wilson, R.S., eds. "Moonlab." Stanford/Ames Summer Faculty Workshop in Engineering Systems Design, June 24–Sept. 6, 1968. NASA Contractor Report NASA CR-73342, 1969.

Libassi, Paul T. "Space to Grow." *The Sciences*, July 8, 1974.

Marvyama, Magoroh, and Harkins, A., eds. *Cultures Beyond Earth.* Vintage Press, 1975.

Maruyama, Magoroh. "Designing a Space Community." *The Futurist*, 104–121, April 1976.

———"Diversity, Survival Value, and Enrichment: Design Principles for Extraterrestrial Communities." *Space Manufacturing Facilities (Space Colonies)*, ed. Jerry Grey. AIAA Publication, March 1, 1977.

———"Extraterrestrial Community Design: Psychological and Cultural Considerations." *Cybernetics* 19, Belgium, 1976.

———"Social and Political Interactions among Extraterrestrial Human Communities: Contrasting Models." *Technological Forecasting and Social Change* 9: 349–360, 1976.

Matloff, Gregory L. "Utilization of O'Neill's Model I Lagrange Point Colony as an Interstellar Ark." *J. Brit. Interplanetary Soc.* 29, no. 12: 775–785, December 1976.

Michaud, M.A.G. "Escaping the Limits to Growth." *Spaceflight*, April 1975.

———"Spaceflight, Colonization and Independence: A Synthesis: Part One: Expanding the Human Biosphere." *J. Brit. Interplanetary Soc.* 30, no. 3, March 1977.

Mirocha, John W. "Space Colonization and Intermediate Technology: Toward a Social Policy." *Futurics* 1, no. 4: 124–130, Spring 1977.

NASA/ASEE Systems Design Institute. "Design of a Lunar Colony." NASA Grant NGT 44-005-114, September 1972.

Nishioka, Kenji; Arno, Roger D.; Alexander, Arthur D.; and Slye, Robert E. "Feasibility of Mining Lunar Resources for Earth Use: Circa 2000 A.D." Vol. I and II. NASA Ames Research Center, NASA TMX-62267, TMX-62268, 1973.

North American Rockwell. "Lunar Base Synthesis Study." NASA Contract NAS8-26145, SD71-477, 1971.

————"Orbiting Lunar Station (OLS), Phase A Feasibility Study."
NASA Contract NAS9-10924, MSC-92687, April 1971.

O'Leary, Brian T. "Mining the Apollo and Amor Asteroids." *Science,*
in press.

————"Space Manufacturing, Satellite Solar Power, and the Human
Prospect." *Futurics* 1, no. 3: 54–60, Winter 1976.

O'Neill, Gerald K. "Colonies in Orbit." *The New York Times Magazine,* 10–11, 25–29, January 18, 1976.

————"The Colonization of Space." *Physics Today* 27: 32–40, September 1974.

————"Engineering a Space Manufacturing Center." *Astronautics &
Aeronautics* 14, no. 10, October 1976.

————"Future Space Programs 1975." Testimony in U.S. Congress,
House Committee on Science and Technology. Government
Printing Office, 1975.

————"The High Frontier," and associated articles. *The CoEvolu-
tionary Quarterly* 7, 6–29, Fall 1975.

————*The High Frontier: Human Colonies in Space.* William Morrow
and Co., 1977.

————"Power Satellite Construction from Lunar Surface Materials."
Testimony before Senate Committee on Aeronautical and Space
Sciences, January 1976. Government Printing Office, 1976.

————"Progress toward Space Manufacturing." *Astronautics &
Aeronautics,* October 1976.

————"Settlers in Space." Science Year, World Book Science
Annual—1976. Field Enterprises Educational Co., 1975.

————"Space Colonies and Energy Supply to the Earth." *Science* 190:
943–947, December 5, 1975.

————"Space Colonies: The High Frontier." *The Futurist,* 25–33,
February 1976.

Paine, Thomas. "Colonies in Space." *Time Magazine,* June 3, 1974.

————"Humanity Unlimited." *Newsweek Magazine,* August 25,
1975.

Parker, P.J. "Lagrange Point Space Colonies." *Spaceflight,* July 1975.

————"Space Colonies." *J. Brit. Interplanetary Soc.* 29, no. 12: 769–
774, December 1976.

Parkinson, R.C. "The Colonization of Space." *Spaceflight* 17, March
1975.

————"The Resources of the Solar System." *Spaceflight,* April 1975.

————"Takeoff Point for a Lunar Colony." *Spaceflight*, September 1974.

Reis, Richard. "Colonization of Space." *Mercury*, 3–4, July 8, 1974.

Richards, I.R., and Parker, P.J. "Estimates of Crop Areas for Large Space Colonies." *J. Brit. Interplanetary Soc.* 29, no. 12: 769–774, December 1976.

Robinson, George S. "Legal Problems of Sustaining Manned Space Flights, Space Stations, and Lunar Communities through Private Initiative and Non-Public Funding." *International Lawyer*, 455–475, 1973.

————*Living in Outer Space*. Public Affairs Press, 1975.

————"Scientific Renaissance of Legal Theory: The Manned Orbiting Space Station as a Contemporary Workshop." *International Lawyer*, 20–40, 1974.

Rosenberg, S.D.; Futer, G.A.; and Miller, F.E. "Manufacture of Oxygen from Lunar Materials." *Annals of N.Y. Academy of Science* 123: 1106–1122, 1965.

Salkeld, Robert. "Space Colonization Now?" *Astronautics & Aeronautics*, 30–34, September 1975.

Sanders, A.P. "Extraterrestrial Consumables Production and Utilization." NASA Report TMX-58087, MSC-06816, 1972.

Smith, David B.D., ed. "A Systems Design for a Prototype Space Colony." Student Project in Systems Engineering, MIT, Spring 1976.

"Space Manufacturing." Collected articles and papers from the 1976 NASA Ames/OAST Summer Study, June 16–July 30. NASA/OAST, Washington, D.C.

Vajk, J. Peter, "The Impact of Space Colonization on World Dynamics." *Technological Forecasting and Social Change* 9: 361–399, 1976.

Vondrak, R. "Creation of an Artificial Lunar Atmosphere." *Nature*, March 1974.

von Puttkamer, Jesco. "Developing Space Occupancy: Perspectives on NASA Future Space Program Planning." *Space Manufacturing Facilities (Space Colonies)*, ed. Jerry Grey. AIAA publication, March 1, 1977.

Index